# Crusaders of Chemistry

# CRUSADERS OF CHEMISTRY

## SIX MAKERS OF THE MODERN WORLD

BY

Jonathan Norton Leonard

AUTHOR OF *Loki: The Life of Charles Proteus Steinmetz*

Garden City Publishing Co., Inc.

New York

PRINTED AT THE *Country Life Press*, GARDEN CITY, N. Y., U. S. A.

CL

1938

GARDEN CITY PUBLISHING CO., INC.

# FOREWORD

THIS book does not pretend to be a complete and inclusive history of chemistry. If it were, it should have five volumes and be written in German. Nor is it a condensation of such a book. It is a study of six men whose lives are red-letter pages in the history of the science.

No monk-like laboratory workers were these six. They were not interested in chemistry alone. Religion, politics, and medicine were not outside their field, for in those days a "natural philosopher" could touch on every human interest without unduly scattering his attention. But each man left a deep and lasting impression on the science we now call chemistry.

All the six men treated in this book lived in times of intellectual crisis. The course of their lives was not smooth. They had to fight valiantly against the prejudice and ignorance of the past, and they knew the exaltation which comes only to the pioneer, the prophet of a new doctrine—for they were prophets of the scientific method, the doctrine which now dominates the world which it has done so much to create.

# Contents

# Illustrations

# The Battlefield

# CHAPTER I

## The Battlefield

And evermore, where that ever they goon
Man may hem knowe by smell of brimstoon;
For all the world they stinken as a goot;
Her savor is so rammish and so hoot,
That though a man from hem a myle be,
The savour wol infecte him, trusteth me.
—*Chaucer,* THE CHANOUNS YEMANNES TALE.

THE chemical laboratory still smells much the same as it did in Chaucer's day, and chemistry is still called "stinks" in the English public schools, but since then the science of chemistry has come very much up in the world. The chemists are no longer hunted from town to town by indignant priests backed up by direct-action constables. Kings no longer hire them in secret to adulterate the currency with synthetic gold and burn them in public when they fail. They no longer bury their textbooks in the earth and consult them by the light of the moon.

Chemistry has a long and checkered history. It

has been practiced by all sorts of men from Egyptian priests to modern college professors. During the Middle Ages it was called "alchemy" and existed as a curious hybrid, part science, part religion, part magic. The alchemists used many of the processes we use to-day, but were apt to throw in a few prayers for good measure. As the Middle Ages passed into the Renaissance, chemistry changed to fit the times, gradually shook off its mysticism, and was ready by the middle of the Seventeenth Century to lead the world toward modern scientific civilization.

In our modern world chemistry occupies a key position. Without it our civilization would fall to the ground. Wheels would cease to turn, crops would fail, and plagues would reduce the population to a sickly remnant. Our automobiles would still be wooden ox-carts if chemistry hadn't learned how to make cheap and strong metals. And the bright colors of modern life would be poor dull things if chemistry had not made the black coal-tar blossom like the rose.

But the battle wasn't an easy one. Not only were the secrets of Nature hard to learn, but the chemists had to fight for the right to discover these secrets and proclaim them to the world. They had to fight the superstitious who considered their work mere trafficking with the devil. They had to fight the

Church which tried manfully to destroy its growing rival, Science, before it got too strong. They had to fight sentimental military men who feared another invention like the gunpowder which destroyed so unfeelingly their beautiful armored knights. And they had to fight against the universal feeling of the majority of mankind that things are best as they are and should be let alone.

Chemistry, like all other sciences, began not as a science but as a series of practical discoveries. The first chemist was the hairy person who discovered that the coals of a lightning-set forest fire could be fed with dry branches, kept alive, and domesticated. The other sciences started the same way. The first physicist was the ape-man who found that he could hit harder with a club than with his bare fist. The first biologist was the primitive herdsman who discovered that two goats, properly introduced, make three.

These are empirical discoveries, not true science. Men knew that certain things happened. They even knew how to control their happening. But they didn't know *why* they happened. And from the very beginning they were curious. Weird and elaborate were the theories they built up. Fire they considered a god who lived on dry wood and sometimes, to punish man for his various misdeeds, consumed whole forests. Lightning was the spear of the Lord. Good spirits inhabited their tools and their growing grain. Little devils helped the frost to split rocks, and greater devils breathed the plague into the nostrils of mankind.

Gradually men learned the habits of these devils and spirits. They learned, for instance, that the iron-making spirit was pleased if a little chalk or lime-stone were added to his diet of ore and charcoal. They found that the fire spirit couldn't get along without the spirit of the air. Learning the mysterious ways of the spirits of material things was the first step toward a practical technology.

Strangely enough this mystical phase of chemistry has not entirely passed yet, as anyone can see for himself if he takes the trouble to go to a large public library and watch the curious people who read the books on alchemy. No one studies mechanics with a prayer book in one hand. Practically no one ever did. The mathematician doesn't chant an ancient charm as he integrates a differential equation. But chemistry is different. There are still plenty of people who be-lieve that certain chemical reactions are not re-sponsible to the laws of the physical universe. There were more in past ages, but some remain.

The reason for this extraordinary vitality of the mystical aspect of chemistry is easy to find. With the

AN EARLY ALCHEMICAL HINT

"He who does not understand this picture does not understand the preparation of the Art."—from "SPLENDOR SOLIS."

exception of biology, which only recently passed the simple-mindedly empirical stage, chemistry was the most complicated and baffling of the early sciences. Its fundamental laws were hidden under many layers of confusing mystery. When the early mathematician discovered that the square of the hypotenuse equaled the sum of the squares of the other two sides, he had found a fact which was completely true. It was part of the mechanism of the universe. He could develop it and use it to find more facts. But when the early chemist found that a certain green stone gave copper if heated with charcoal, he knew no more about the mechanism of matter than he did before. Other green stones gave no copper. Some black stones did. The best the chemists could do was to build up a mass of practical information which enabled them to get copper as easily as possible and hope that some time in the future they would know the *why* as well as the *how*.

This sort of practical information began to accumulate very early. The Egyptians had a great store of it, which they used with much skill. Their colored glass was as good in many respects as the best we can make now. They must have had very fine steel or they would not have been able to perform their marvelous carvings on the hardest materials. The cement of the pyramids has defied the weather of four thousand years, something which modern cement is very unlikely to do.

But there they stopped. They never worked out any valid theories to explain the reactions they used so successfully. And neither did the Hindus, the Greeks, or the Romans. The stock of empirical knowledge was passed down through the ages. Some of it was lost, and some new methods and processes were discovered. But whenever the more contemplative of the chemists tried to reason out the rules of their science, they came sharply against a blank wall. They couldn't establish a single fundamental law. The subject was too complicated. They were up against one of those terrible problems, no part of which can be solved without solving very nearly the whole.

This is the way it worked. Take a comparatively simple reaction, such as the hardening of lime plaster when exposed to air. It hardened all right. Men had observed this from the earliest times. But no one knew why. We now know that the carbon dioxide in the air combines with the calcium hydroxide of the plaster to form calcium carbonate and water. But the ancients didn't even know there was such a thing as carbon dioxide. And they thought with good reason that water was the simplest and most elementary of all substances. Or take the burning of

wood. We know that the oxygen of the air combines
with the carbon and hydrogen of the wood to form
carbon dioxide and water. But the ancients didn't
dream of oxygen. They had no reason to think that
air was not a simple substance, and to discover that
it was not, they would have had to develop balances
capable of weighing minute quantities; they would
have had to make apparatus to control elusive gases,
and they'd have had to learn to shut their eyes to all
the side manifestations so conspicuous in combustion
—the flame, the smoke, and the ash. There was no
chemical problem which could be isolated and
studied by itself. You had to grasp the whole subject
or nothing.

Confronted by this apparently insoluble mystery,
the ancient chemists did what might have been expected. They concocted sweeping theories about
matter and tried to prove that the facts agreed with
them. Some of these theories are quite reasonable.
Aristotle, whose theory was the best of the lot, decided that the universe was made of four "elements,"
earth, air, fire, and water, and this theory, although
obviously imperfect, was the best man could do for
two thousand years. The work of the practical
chemists went on; they discovered many isolated
facts. But until the Eighteenth Century, the essential

mystery was hardly better understood than it was in the days of Alexander the Great.

If all men had been like Aristotle, the course of science would have run smoothly. Men would have worked patiently gathering facts, stored them away, and waited until they had accumulated enough of the pieces to put together into a perfect pattern. But that's not the way of human nature. It's human to read the last chapter of a novel. It's human to generalize beyond the facts. Also it is human to invoke divine aid to guide your researches.

Early chemists were very seldom philosophers like Aristotle. They were men of the people, trying dimly to cope with a subject which they did not understand. And just as the Egyptian priests invented gods and demons to explain the reactions in their furnaces, the chemists of the Middle Ages attributed to God the powers which they could not prove were inherent in the materials they worked with.

This sounds harmless, even quaint, but it's one of the unfortunate things about human nature that as soon as religious ideas take a hand in a problem, the facts have to retire to a back seat. Apparently most human beings need some sort of religion, something to believe without proof. But the tragedy occurs when an article of faith devised by theological minds to explain an unsolved problem declines to be replaced by a subsequent and rational solution of that problem. But this happens wherever religion and reason exist side by side. Religion is the *status quo*. It is the scrap basket into which are thrown all problems which cannot be solved by other means. And when a man comes along who can, with unaided mind, dig deeply into the mass of natural fact and unearth a reasoned conclusion, all the bureaucrats of religion—the priests, the theologians, the magnates of the Established Church—pounce on him like a pack of wolves. And they do well to do so. For in every active epoch of human history religion has been like a sandy island eaten away by the sea. It has no hope of gaining territory. It must defend what it has. And the man of science with a new explanation for one of the mysteries of Nature is an enemy to be feared and combated, for if he explains, for instance, that lightning is an electrical discharge, not a weapon of the Lord, the Church as well as the Lord has lost a stronghold.

There'll be a good deal about religion in the following chapters, for the chemists up to the end of the Eighteenth Century had to fight with the Church for every inch of the ground. The Church has long since abandoned its positions against chemistry. The battle is over, and religion is throwing its remaining strength against the newer and more vulnerable

sciences of biology and psychology, but chemistry
for centuries bore the brunt of the battle. In its
alchemical phase it was treated by the Church as
just another heresy. After Paracelsus it had to fight
the battles of medicine, a profession of which the
Church has always tacitly hoped to take control.
And later when the scientific method became established, chemistry, at that time merged with physics,
was the first science to make it bear definite fruit,
and the Church very rightly regarded the Scientific
Method as its most dangerous enemy.

So it was no accident that many of the early chemists were as much concerned with religion as they
were with science. Roger Bacon was a friar who
hoped to establish a new system of thought based on
reason rather than on faith. The Church imprisoned
him. Paracelsus tried to take medicine from the
power of the Church and the Galenic physicians who
were tolerated by the Church because their information came from the same source as Christian theology.
The Church drove him into vagabondage. Robert
Boyle was an intensely pious man, but his champion-
ship of the scientific method brought him under attack, and he too would have fallen if he hadn't been
a nobleman powerful at court. Joseph Priestley was
a Dissenting minister whose scientific doctrines led
him to advocate Unitarianism, which, since it is half

religion and half reason, the Church hates both as an enemy and as a traitor. He was exiled. Cavendish escaped only because he was a very rich man who desired no contact with society. Not until we come to Lavoisier do we find a scientist who didn't have to battle with the Church. And he, poor man, came in conflict with a different if just as ferocious force, the French Reign of Terror.

The battle is over now, and Chemistry with certain of the other sciences is something of an established church itself. It has its endowments, its institutions. When rich men die, they leave it large sums in the same spirit in which they once endowed the monasteries, churches, and chantries. After-dinner speakers give it credit for increasing the happiness and well-being of humanity, and they sound quite like their medieval equivalents who thanked Christianity for these benefits.

But the story will repeat itself. It always does. Perhaps the Christian Church will never regain enough strength to go on a crusade against science, but another *status quo* will establish itself and defend its doctrines against all comers. Perhaps science itself will become a conservative dogma desperately and ferociously opposed to a new kind of thought growing out of modern philosophical physics or the still-to-be-discovered phenomena of the mind. Then we'll have the story of Bacon, Boyle, and Priestley all over again. Audacious and earnest innovators will battle with equally earnest conservatives who want the world to stand still. Then someone, late in the cycle, will look back and find to his surprise that men before him had to fight for the right to reason, for the right to think for themselves.

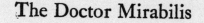

# The Doctor Mirabilis

# CHAPTER II
## The Doctor Mirabilis

I T WAS an April morning in the year 1268, and the sun had just risen. Past the open end of the courtyard ran the Seine, high with spring rain and dark blue in the morning light. The towers of Notre Dame cast jagged shadows across it, and a snub-nosed riverboat floated down noiselessly, bound for Havre and loaded with French wines for England.

From a door under the cloister came a tall, middle-aged man dressed in a brown robe with a brown rope around his middle. His head was bare and shaved, and a rosary of wooden beads hung around his neck. He stretched a little, rubbed his eyes, looked at the river and the boat, looked up into the blue sky where a small white cloud was chasing a still smaller one, and turned squarely to the east to let the sunlight clear the shadows out of his mind. Then he crossed a corner of the courtyard, opened another door, and entered the strangest room in Christendom. It was his laboratory—the laboratory of Roger Bacon.

The room was large, but from end to end it was filled with the weirdest collection of curiosities ever assembled by a Christian scholar. The walls were covered with books, some bright and new, some black with the fingering of ages, some of clumsy parchment, some of soft, porous paper. Among them were rolls of brown papyrus and great folio sheets pressed between wooden boards. Other shelves were filled with minerals glittering in the yellow sunlight or dull with the dust of neglect. From the beams overhead hung bunches of dried herbs, small leather bags, barbaric weapons with inscriptions in strange letters, and other things to which Bacon alone could have put a name.

Littered about on tables were geometrical diagrams, compasses, rulers, and quill pens of various sizes. A large charcoal furnace sat solidly in a corner, and near it were racks of glass and clay vessels— alembics, crucibles, and curcurbits. Heavy oak cupboards flaunted large keyholes and gave the impression of being very securely locked.

Bacon sat down before an upright desk and opened a manuscript book in some graceful Oriental script. A knock came at the door, and a young man entered.

"There's a brown friar at the outer gate, master," he said. "I told him it was too early in the morning, but I couldn't get rid of him."

"Who is he?"

"Brother Manfred."

"Throw him out, the sneaking spy."

"But Master, you know what he's like. He'll tell everyone in town you didn't dare see him."

"He'll be right too. I'm afraid of him. His hobby is heresy. He could find it in the works of Lord Jesus himself. But show him in. Wait though. I've got some things here I don't want him to see."

Bacon went to a table and picked up an elaborate contraption of brass and small lenses. He put it into a cupboard and tucked some loose sheets of paper in beside it. He locked the cupboard with a large key.

"All right. Now we're ready for him."

Presently Brother Manfred arrived, smiling a rather unconvincing smile. He was a small thin man whose friar's robe flapped loosely about him. Bacon bowed with poorly concealed hostility.

"It's a great pleasure," he said coldly, "to receive a visit from so learned a man so early in the morning. Most of the schoolmen are busy with their lectures this time of day and haven't any time for poor tinkerers like me."

"You're not a tinkerer, Brother Roger," replied the friar. "Your learning is the pride of our order. They say you know more than any other man alive."

"I know some things, Brother," said Bacon, "but only those things which diligent study can teach. And you, my friend, are famous for diligence."

Brother Manfred went across the room and picked up the book which Bacon had been reading.

"What's this?" he asked. "The letters are skillfully made, but I can't read them."

"That's a holy book, written by a Christian of Damascus."

"Is it heresy?"

"Perhaps. But in those days heresy was not as inexcusable as it is now. The Blessed Apostles walked in the streets of Damascus, and the writer of this book may have talked with Our Lord himself."

"True, true," said Brother Manfred. He went to a table and picked up a large white flower, somewhat faded, which lay there.

"You seem to love flowers, Brother Roger. But you've pulled one of the petals off this one."

"Our Lord Himself said 'Consider the lilies of the field.' He knew without looking how the petals were attached, but poor mortals like us have to find out for ourselves."

Brother Manfred went to the other end of the room where the furnace stood among its racks of apparatus.

"An alchemist would be glad of this array."

"Pope Sylvester the Second was an alchemist who learned his art among the infidels of Spain. Yet he lived to hold the keys of heaven."

Brother Manfred turned away from the furnace and faced Bacon squarely.

"They tell me," he said, "that you have an engine made of crystal which lets you look into the soul itself. I should like to see it."

"You are mistaken. There is no such engine here."

"I heard it on good authority. Perhaps you wouldn't know the soul even if you saw it plainly."

"Perhaps not. I've taken only minor orders."

Manfred went slowly to the door.

"Now I must go," he said reluctantly, "but some other time I should like to come and hear more of your work. Your learning is the talk of Paris."

"You will be welcome. I shall be glad to show you the few poor truths I have found in the works of the Fathers."

As soon as the door closed behind Manfred, Bacon winked happily at his young assistant.

"How were my answers, Peter? Did I look nervous?"

"You did fine, Master. I haven't heard such piety since I came here."

"When I was young, I argued in the schools and learned all the little tricks they teach there. It's a silly business, but it came in handy to-day. It's good he couldn't read my book or look through the door of my closet."

"He was as likely to look through the door as he was to read the book. The friars know a lot about the Scriptures and the Fathers, but that's all they do know."

"Right, Peter. You know the friars as well as I do. But now we'll have to get to work. You make two copies of this letter. In half an hour John will be here all ready to start."

Peter went to a table and began to write. Bacon sat down at another on which lay two large books bound in bright new leather, marked "Volume I" and "Volume II." Bacon placed them one upon the other, turned both backs toward him, and read the names on the spines: *"Opus Majus,* Volume I, Volume II, by Roger Bacon." He did this slowly as if his mind were far away. Then suddenly he aroused himself and with a few quick motions wrapped the books in a square of cloth and put them in a leather case made to fit. He turned down the heavy cover and strapped it tight. Then he leaned back in his chair, put his hand over his eyes, and waited.

After a few minutes the door of the room opened and a young man, hardly more than twenty, appeared against the sunlight. He wore a traveling costume and carried a partially filled knapsack on his back.

"Here I am, Master," he said gayly, "all dressed and ready to go."

Bacon looked up from his reverie. His whole expression had changed. Not the shadow of a smile hid among his features and he looked years older. He weighed the case of books in one hand; then laid it down on the table.

"John," he said slowly, "you were my first pupil, and the only one I've educated according to my own ideas. You can think in straight lines, not in neat circles like the best of the schoolmen, or in aimless snarls like the worst of them. And now——" Bacon picked up the case of books and slammed it down on the table with a crash. *"Now,* do you realize what you're taking to Rome?"

"I ought to," said John lightly. "We've worked on it long enough together, you bringing great thoughts out of your head and I trying to understand what they're all about."

"You understand well enough," said Bacon, "better than the schoolmen who strut around Paris so pompously. But that's not what I mean. Come nearer."

John sat down on the bench near the master.

"That leather case," said Bacon slowly, "is the

most precious load in Christendom. I'm not speaking lightly when I say this. I mean it. If the knowledge it contains and the theory of thought it teaches should become accepted throughout Europe, they will change the aspect of the whole world. We Latins are now a poor fringe on the outskirts of civilization. The Saracens know more than we do; the Turks throw back our armies; and out beyond the Black Sea rage the Tartars, millions and millions of them, savage, without heart, mind, or conscience, whom an ill wind might any day blow down upon Europe. Never did our race hold more precariously to life.

"Our leaders do little to guard against these dangers. Our soldiers fight well, deal their blows bravely, and die with the name of Christ on their lips. But they accomplish nothing. The Turks and Tartars are brave too. They've got as good swords and better horses. They'll never be conquered by stupid force. There are too many of them.

"We Latins have one weapon which can save both us and Christianity. It's rusty with neglect. The schoolmen have put it away in the corners of their minds and take it out only from curiosity. This weapon is Science. If we polish it and sharpen it again, we shall be irresistible. We shall extend Christianity to Farther India, to Cathay, right round the world. No Tartar, no Turk or Saracen will stand against us. The Pope won't shiver in his slippers every time a messenger comes from the East."

"I know this," said John, "and I know how important it is. I won't drink a drop of wine or speak to a single girl until I've put the books into the Pope's hands."

"Oh, you can resist temptation all right, but that won't be enough. Your job isn't that of a mere messenger. You're an ambassador from a country which as yet has only a handful of citizens. And such an ambassador has to work carefully.

"The Pope is a wise and broad-minded man, but after all a Pope is a Pope, and they all have to have certain prejudices. That's their job. They guard the established order. They defend the faith against all comers. They rule temporal Christendom as much as they dare, and watch closely for any tendency which may lessen their authority. The ideas in my book will do just that. They will raise up a power much stronger than faith, and perhaps the Pope and all he stands for will go down before them. So you see you'll have to be very careful.

"This book of mine is planned so as not to arouse the Pope's suspicions. I've told only the truths which won't alarm him. I've avoided heresy as much as I could. I've promised great benefits for Christendom and I've proved as well as I could that the study of science will lead to a more perfect understanding of God through knowledge of His works.

"But I don't hope to deceive His Holiness *completely*. That's not possible, and it's not necessary. If he can be persuaded that he can reap the advantages of my program without doing too much violence to the fabric of the Christian faith, he'll come around to my point of view. Don't lay all your cards on the table. Remember that as you go up the scale of power, the methods of diplomacy become more and more insincere. The Pope is at the very top. He can't be frank and open, or he wouldn't be Pope. So don't tell him all you know. He won't be able to see by himself the ultimate results of our program, and you mustn't give him any hints. Demonstrate the immediate advantages of our system, and don't discuss such things as the faith or the Scriptures. That's his specialty, and he can do it better than you can. Be simple-spoken, or appear to be. Offer him a weapon against the enemies of the Church and make him realize the power of it. Don't let him suspect that far in the future the weapon may kill the Church that wields it. Do you understand?"

"Yes, I do, Master. We've talked about this before, and you've taught me what to say."

"I know I have, but I wanted to go over it again so it would be fresh in your mind. I think you know how to act. Now for the tricky business of getting the books before His Holiness."

"The road's pretty safe now, and I'll have the sanctity of a pilgrim."

"That's not the point. These books must reach the Pope himself and not be fended away by any of the cardinals, monks, and friars who surround him like hornets around a drop of honey. The trouble will begin the minute you get to Rome. You'll travel as a simple pilgrim. Remain one when you get there. Don't stay at the house of my order. Don't tell anyone what you carry. Don't talk at all. Go direct to the palace and present your letter. I'll give you two copies in case one is intercepted by my enemies. The Pope will grant you an interview."

"And I'll give the books into his own hand."

"By all means. But even then your job will be only half done. Not only must the book reach the Pope, but it must be guided safely into the inner depths of his mind. His courtiers will offer to read the books for him—to save him time and trouble. But he must read them himself. And you must see that he does. Tell him what the books are and what they contain. Remind him that he asked to see them. Then ask for a chance to explain the obscure passages."

"I'll do all this, Master. I know what it means to you. I'll think about it all the way to Rome."

From the street outside came shouts and the sound of horses' hoofs.

"I must go, Master," said John. "Here are my Norman pilgrims. They want to get started early."

Bacon followed him into the courtyard and watched him into the street. The pilgrims' noises died away in the distance, but Bacon still watched, his hopes reaching out toward Rome.

★        ★        ★

It was a small matter, this God-speeding of a manuscript. There were no academic fanfares. None knew of its departure; few even of its preparation. If the great philosophers of the day, Albertus Magnus or Thomas Aquinas, had looked into it, they would have turned up their noses patronizingly, or perhaps taken the trouble to denounce in a few thundering words the damnable heresy it contained. The teachings of Bacon had little vogue in the schools of Paris. They were too earthy, too eyes-to-the-ground; they led too little toward "a perfect understanding of God" to please the theological acrobats who contorted themselves skillfully in a self-made lather of words. The schoolmen of Paris smiled at Roger Bacon with his mirrors, his furnaces, and his habit of talking eagerly with anyone, knight or peddler, who'd visited strange lands. "What's this got to do with salvation?" asked the schoolmen. "What has the rainbow to do with philosophy except to show the glory of God?" They waived him aside as a mere meddler in matters too small to notice. And the little boys of the streets would steal into the courtyard and look through the windows of the laboratory where Bacon was busy with crucibles, diagrams, or curved lenses of rock crystal. They'd wonder at the sparks and the glowing metal, then run away laughing when Friar Bacon stood up and glared at them across a bright bed of charcoal.

What sort of man was this "Friar Bacon" whose labors were so slightly esteemed at Paris? His own age knew little about him, and this little was forgotten soon after his death. A curious legend grew up in its place. "Friar Bacon" became a magician who made brazen heads speak, cities open their gates to besiegers, and cavalry fly through the air. He signed contracts with the devil, who appeared as a beautiful woman shining white among the coals of his furnace. He could walk into stone walls and disappear. The secrets of the king's council room and bedchamber were known to him by magic. He could make gold from mercury, lead, or the tarnished brass of a cooking vessel. Like many a scientist to follow, he became a warlock because the public of his day, unable to understand him, preferred to think he could not be understood at all.

Roger Bacon was born of good family at Ilchester, Somerset, about the year 1214. Almost nothing at all is known about his childhood. When still a young boy according to modern reckoning, he went to Oxford and came under the influence of Robert Grosseteste, who taught in the school of the Franciscans from 1230 to 1235.

This Robert Grosseteste was a remarkable man who might be called the leading liberal thinker of his time. His influence made the intellectual atmosphere at Oxford decidedly less murky than at Paris, where such scholars as Alexander of Hales were already building up the imposing but unsound structure of Scholasticism. Grosseteste was something of a mathematician, something of a scientist. He was interested in various sorts of learning which were much disapproved by his fellow churchmen. He even went so far as to bring several Greek scholars from schismatic Constantinople to visit him at Oxford where they left behind them a tenuous tradition of Greek learning. After he became Bishop of Lincoln, the administrative affairs of his important diocese occupied more of his mind, but while at Oxford he was a landmark of common sense and enlightened learning in the hopeless obscurity of medieval thought. His influence started Bacon on the difficult road which led to what he called "the True Knowledge."

There is no use trying to give an accurate account of the details of Bacon's life or his movements between Paris and Oxford. Every date has to be followed by a question mark. He probably traveled frequently between the two. Scholars in that day were citizens of no country. They spoke Latin, lived in the houses of the Church, and traveled about to hear famous teachers or to give lectures of their own. Let it suffice that Bacon earned an Oxford degree of M. A. and won much renown in the schools of both countries as a lecturer on Aristotle. He also became a Franciscan friar, probably to enjoy the protection of this order and use its excellent facilities for study. In the Thirteenth Century the Dominicans and the Franciscans between them had gained such a dominant position in the universities that not to belong to one or the other was a severe handicap to any scholar. The myth-makers who spun stories about Bacon took the fact that he was a "brown friar" and pictured him as wandering about the country, begging at manors and inns, and generally practising the various chicaneries for which the friars were famous. This was far from true. Bacon lived almost his entire life at Paris and Oxford, and devoted himself exclusively to study and research.

The Thirteenth was one of the great centuries of history. It marked the high point of medieval civilization, and in many ways has not been surpassed by any succeeding period. The cathedrals still defy imitation. The songs of the troubadour and minnesinger still enchant those few who can read them. And no man since Thomas Aquinas has constructed a system of thought perfect enough to be accepted by two hundred million people after six hundred years. Thirteenth Century scholars were as subtle as any to-day. Society was fairly well organized and largely contented. But the Thirteenth Century lacked one thing: a program for the future.

If a broad-minded modern scholar were to turn the clock back some seven hundred years and find himself in Paris at the middle of the Thirteenth Century, his first sensation would be delight at the intensity of intellectual activity. The scholars were the men of the hour. The Church governed most of the activity of the city, and its leaders were men whose thoughts were among the stars. The most promising road to fame led through the university, and this road was open to all. From the ends of Europe they gathered. If a monastery, cathedral school, or the court of a distant noble produced a young man with promising intellect, he'd be provided with funds, books, and letters of introduction and shipped off to Paris for the greater glory of learning and the reputation of his province. The political conflicts of the feudal world were heard at Paris as only a distant rumble. The scholars walked two by two along the banks of the Seine, thinking high thoughts and living in a paradise of intellectual detachment.

But our transplanted modern would soon feel a second and not so reassuring sensation. Where was this feverish activity leading? What did these brilliant churchmen and friars really hope to accomplish? Was their program continuous? Were they working at a problem with unlimited possibilities which would yield valuable results from time to time? Unfortunately they were not. Medieval scholarship was like a cavalcade, brilliant, colorful, and gay, but headed for a blank wall. Like a mathematical quantity which approaches a certain value but never reaches it or goes beyond. Like $\pi$ — 3.1416—which gets nearer and nearer to the ratio of the circumference to the diameter, but never reaches it and never even hopes to become 4. Scholasticism was a backward-looking system dependent on the authority of the past, and as such was self-limited in range. The Scriptures and the writings of the Church Fathers contained only a certain number of statements. These could, with great labor, be arranged and reconciled. The tremendous mind of Thomas Aquinas did this. There was no more to do, and Scholasticism smothered in a quicksand of meaningless words. It took the twin explosions of the Renaissance and the Reformation to break the steel network which Scholasticism had woven about the mind.

"What fools they were," says the thoughtless modern whose mind has been trained from childhood to think along more or less scientific lines. "What blind and silly fools. Why didn't they get out in the sunlight and investigate some of the things in the world around them? Why didn't they take the horn of a sea-unicorn and see if it really burst into perspiration when touched with poison? Why didn't they get a diamond and see if it really split to bits when dipped in goat's blood?" Well, the diamond merchants probably had, and Roger Bacon did, but such experiments were contrary to the spirit of the time, unworthy of serious and pious men. "The life on earth is nothing. Salvation alone matters at all. The Scriptures contain the information on this matter. The Church Fathers have added illuminating explanations. What can be more valuable than the study of these two sources? Physics, medicine, astronomy, what are these little things compared to the Perfect Understanding of God? They are only of this life. What are three-score years and ten compared to eternity?" So said the schoolmen. So said the Church. So said the kings and the nobles. Only Roger Bacon and the obscure artisans in their workshops thought otherwise.

Such habits of thought, silly as they may seem, are not mere foolishness. They involve problems of fundamental human values which science, even modern science, hasn't even attempted to solve. If "salvation" is possible and as desirable as the medieval schoolmen thought, and if it may be attained by knowledge of the Scriptures, then certainly the silly laws of the physical universe are exceedingly unimportant by comparison. So said the schoolmen, and they were right as far as they went. A change in their intellectual aims would involve a change of faith, and faith they considered their most precious possession. One man, however, saw the blank wall looming up ahead of the procession. This man was Roger Bacon.

A serious man was Bacon, a master of the dialectic sleight-of-hand so much admired at the time, a reader of strange languages and curious about distant countries beyond the borders of Christendom. In these ways he was not unique. In one respect only he differed from the men of his time. He had little regard for authority. When he read in some ancient

author that a vessel of hot water freezes faster than
one of cold, he didn't accept this as ultimate truth.
He took two vessels exactly alike, filled them with
hot and cold water, and set them outside in the
street. When the cold water froze first, he didn't
conclude that his eyes deceived him or that the devil
was laughing down the chimney. He said the ancient
author was mistaken or a liar. When a diamond-
cutter told him that he broke diamonds in a mortar
like anything else, he didn't call him a scoundrel,
but concluded that Pliny knew as little about diamonds as he did about the tides.

In the Thirteenth Century a man in a scientific
state of mind found plenty of work ready at hand.
The frontiers of knowledge had not been pushed
so far away that it took a lifetime to get beyond
the well-mapped territory. A cup of water, a mirror, and a ray of sunlight contained easy truths
which no one in the world had observed before. The
goldsmith round the corner, the cook in the kitchen,
the peasant bringing his vegetables to market knew
isolated facts which an able mind might put together
into a regular pattern of science. Roger Bacon proceeded to do this. It wasn't easy. A thousand years
of Christian philosophy, a thousand years of classical
learning had to be thrown overboard or tested rigorously for error. Long-standing habits of thought

must go. The doctrines of a beloved religion must
be put aside as the Scriptures and works of the saints
were passed under the pitiless lens of skeptical
scrutiny. But the results were amazing.

Gradually the conviction developed in Bacon that
he was born with a tremendous mission. He wasn't
an imaginative person; he didn't call himself a
prophet. But the more he learned about the physical
world around him the more he became convinced
that the scholarship of the Latins was on the wrong
track. That its efforts led nowhere. That the future
of Christianity, even its preservation from the appal-
ling dangers looming up out of Asia, depended on a
new start, a new learning built on something more
substantial than the barren theology which now absorbed its energies.

With this thought in the back of his mind he
forced himself to work as few men had worked before. Eighteen hours of study a day, two of recreation, and four of sleep. His health broke down and
he had to retire from the university. Soon he was
back at his desk. His eyes gave out. He invented
spectacles to correct his failing vision. He planned
his diet carefully—perhaps the first in Europe to
do so. A mass of knowledge accumulated in his head
such as no man had ever possessed before.

But study alone was not enough. Wherever he

turned, he soon exhausted the known facts. He'd have
to find more. So, borrowing ideas for apparatus from
furtive alchemists, from artisans, from cutters of
precious stones, he went to work to test the facts he'd
winnowed from the writings of past ages. His considerable fortune melted away, blown up the chimney
with the sparks of his furnace, paid out for lenses
and curved mirrors, for curiosities from foreign
lands, and for such trade secrets as were not given
free. But the great idea was taking shape.

In 1266, when Bacon was fifty-two years old, his
great opportunity came. Guy de Foulque, a French
scholar, became Pope as Clement IV. He was a
broad-minded man for the period and had heard
from common friends about Bacon's remarkable
work. Perhaps the controversies which Bacon's
"heresy" had stirred up at Paris were appealed to
Rome. At any rate he wrote Bacon to send his works
"in all haste and in secret, notwithstanding the prohibition of any prelate or constitution of his order."

Bacon was tremendously excited and somewhat
terrified. He wrote back to Rome pleading for time.
The matter was too large for quick action. Nothing
was written down in proper form. His Holiness
would please give him a year to put his results together in a book worthy to pass before his supreme
judgment. The respite was granted, and Bacon gath-

ered all his remaining resources and buckled down to the tremendous task of writing an intellectual program for the world.

The great work, the *Opus Majus,* which Bacon and his devoted pupils put together in the eighteen months after receiving Clement's letter, is almost an encyclopedia of medieval knowledge, at least that part of it which had penetrated up from the empirical discoveries of artizans and unlearned travelers. Some things Bacon left out—from prudence —but still the book contains enough to surprise any modern who thinks that a man of the Thirteenth Century couldn't have had the true scientific spirit.

Bacon didn't consider the *Opus Majus* a straight scientific treatise. He called it a *persuasio,* that is, a book written with a purpose, to convince the Pope that the Latin world needed a new intellectual program. Persuasion is not written solely for the reasoning mind; it also considers the sentiments and prejudices of the person addressed. Popes have strong prejudices, or they pretend to have, which amounts to much the same thing. So Bacon, in planning his work, had to restrain his scientific enthusiasm, had to put over as strongly as possible his ideas about the necessity of experimental science and at the same time reassure the Pope that innovations in this direction would help Christianity, not cause its ultimate decline. Bacon's scientific method was the fundamental heresy, the heresy with which the Church still fights a losing battle; his problem was to cloak it in such conciliating words that the Pope would hardly realize it was there.

A difficult task certainly, and the trouble began with the first section, in which Bacon demolishes the intellectual method of the time and shows the causes for its stagnation and practical failure. Many are his circumlocutions, long are his conciliating approaches, but finally he has to get down to what he isolates as the root-cause of error—belief in the authority of the past. Here's where he had to step carefully. The Church draws a distinction between faith and belief in traditional authority, but it's one of those distinctions which don't convince anyone who hasn't a strong will to believe. Bacon pretended to understand the difference perfectly, and he opens his chapter with a long statement to that effect. Then he launches a savage attack on the blind belief in the past which was the evil genius of medieval thought.

"Four causes of error there are, among the Latins," says Bacon. "These are dependence on authority, yielding to established custom, allowing weight to the general opinion, and concealment of real ignorance with pretense of knowledge." The first three are related and boil down to a slavish devotion to the past. The fourth, as Bacon says, is common to all ages and has to be guarded against wherever it appears. Bacon's language is careful and judicious, but through the moderate statements you can feel the fire of his wrath against the conservatism of the schoolmen. He complains bitterly that the method of argument in his day was something like this: "It is affirmed by predecessors, it is the customary view, it is the popular opinion; therefore it must be correct." If anyone disagreed, he was overwhelmed with quotations from the Scriptures, and his only defense was to quote back. Even then the simplest observed truth wasn't sure of winning. Bacon almost says that you can prove anything from the Scriptures if you are clever enough. But he stops just in time. After all he was writing a *persuasio* for the Pope. And the Pope himself wouldn't dare to let such blasphemy go unpunished.

Having dealt with the schoolmen, Bacon proceeds to run over the stock of scientific knowledge of the time, working into it certain of his own discoveries and showing where possible the value to Christianity of increased knowledge in the field. This part of the *Opus Majus* looks rather feeble to our eyes. The sciences of Bacon's day were few in number and limited in extent. Except by the Arabs, almost noth-

ing had been done since Aristotle, and the Arabs
confined themselves largely to translating Aristotle
and commenting upon him. But judged by comparison with the other scientific books current at the
time, Bacon's work stands out as a shining beacon of
scientific rationalism. Its clearness and reasonable-
ness within its field are striking and do Bacon great
credit. Not only did he have to start from the very
bottom, unsupported by a tradition of straight thinking, but he also had to clear away a vast accumulation of doubtful facts which were often hard to test.

When Bacon has no means of testing a statement
rigorously for himself, he falls back on the best
opinion of the past. And this was all he could do
under the circumstances. Modern scientists do the
same. He couldn't journey to the headwaters of the
Nile to see if shadows fall north half the year and
south the other half, so he takes Ptolemy's word for
it since the theory seems to agree with other known
facts. It turned out he was right. He wasn't always
as fortunate, and sometimes he sounds rather silly,
but never does he make an entirely unreasonable
statement about a matter which he could conveniently
observe for himself.

This unavoidable inability of Bacon's to cover the
whole field of science single-handed gives a very
spotty appearance to the central portion of the book.

Some chapters, those on mathematics and optics, for instance, are very good, while those on music, geography, and astronomy, which Bacon considered divisions of mathematics, are rather weak. With Turks, Tartars, and Arabs fighting each other all over the East and slaughtering any Christian they could get their hands on, the conditions did not favor geographical exploration. Music should not be treated as a science at all, and astronomy was so confused with astrology all through the period that even Bacon did not think to separate them.

Bacon's chapter on geography, erroneous as it is, contains in some respects the most interesting ideas in the book. Like everyone else in Europe, Bacon lived under the impending terror of attack from the East. For two centuries the fortunes of Europe had been fading. The Crusaders were driven gradually out of the Holy Land; the Greek Empire seemed trembling at the edge of destruction as fresh multitudes of savage enemies swept down from central Asia or up from the deserts of Arabia. In Bacon's youth Genghis Khan had carried everything before him from China to the Caspian, and his grandson, Batu, had scattered the Poles, defeated the full power of the Germans, and been prevented from ravaging the whole of Europe only by a message which called him back suddenly to the East. In 1258 these same

Mongols had destroyed the Caliphate, leveled Bagdad to the ground, and defeated the Saracens and Turks, against whom all Europe had struggled two centuries in vain. It was a gloomy and uneasy time for Christendom. At any moment a new onslaught might burst from the steppes of Russia. There would be no warning. The Tartars rode well ahead of the news of their coming. Bacon looked across the Seine from his peaceful courtyard and imagined the sack of Paris. Streets littered with bodies; men and women, monks and nuns cut to bits or flayed alive; houses in ashes, and the Tartars encamped in the public squares, their eyes in popular belief dripping small streams of ice-cold blood. Then he'd turn back to his work with new determination. He was the only man in the world with a weapon to make Europe safe from the attack of Asia.

To be sure, the specific weapons with which Bacon hoped to equip the armies of the West were not very formidable. He speaks of a curved mirror which by converging the rays of the sun might set fire to besieged cities. He attributes great power to the Tartars' supposed knowledge of astrology. And although he was apparently familiar with gunpowder, he doesn't think of using it except in firecrackers to frighten the enemy. But Bacon was no soldier and doesn't pretend to know much about military practice. His plea is for an organized campaign to learn the secrets of Nature. When this is done, military devices will appear of themselves:

> "We must consider that although other sciences, such as geometry, do many wonders, yet the wonderful things of most service to the state belong to Experimental Science, which teaches us how to make and use marvellous instruments. The enemies of the Church of God should be destroyed rather by the discoveries of science than by the warlike arms of combatants."

It's astonishing to observe how right he was. All during the Middle Ages, Europe, the small, remote, disunited peninsula, barely maintained itself against the East. Its outside conquests were short-lived, and the Turks, Tartars, and Saracens, when not fighting among themselves, won nearly every conflict. Only after the Renaissance, when Europe had waked up and invented firearms, superior sailing ships, and the better methods of agriculture and industry to keep its armies supplied did it feel safe from Asia, did its conquest of the world begin.

From a scientific standpoint, Bacon's treatment of optics, scattered more or less through the book, is the most impressive thing in it. It's the part which makes us most certain that he didn't put down all he knew. Optics looks like a safe enough subject, and in itself it is. But its two great discoveries, the telescope and the microscope, have done more to change our view of the world around us than all other discoveries put together. The telescope has told us that the earth is not the center of the universe—a very revolutionary conclusion from a theological point of view—and the microscope has shown us the world of the bacteria and proved that even the sacred human body is composed of cells very much like them in structure. A man afraid of the dead hand of the Church would be wise to keep his knowledge of these two devices secret.

It is not certain that Bacon actually constructed a microscope or a telescope, but all through his works are hints, perhaps unconscious, that he had:

"The wonders of refracted vision are even greater. Very large objects can be made to appear very small and the reverse, and very distant objects will seem close at hand and conversely. For we can so shape transparent bodies and arrange them in such a way in respect to our sight and objects of vision that the rays will be bent in any direction we desire, and under any angle we desire we shall see the object near at hand or at a distance. A child might appear a mountain. We might make the sun, moon, and stars in appearance to descend here below."

This chapter is merely a prediction. Bacon says that a device for seeing at a great distance *might* be constructed, and in the previous sections he explains in very scientific language just why this is so. There is no direct statement that he actually had constructed such an instrument. But he had the lenses. They were nothing new. He understood how they worked and why, and a simple astronomical telescope can be made by any child out of two convex lenses arranged at the proper distance from each other. This distance doesn't even have to be accurate. In another passage Bacon remarks that an instrument for bringing distant bodies nearer also "makes the highest parts lowest and vice versa," which is exactly what an astronomical telescope does. The image is reversed. Here is where Bacon gives himself away. It is very unlikely that he would have known this startling fact if he had not actually constructed a telescope and used it.

The situation with the microscope is similar. The simple magnifying glass was perfectly familiar in Bacon's day. Had been for ages. There is evidence that the Egyptians used it. But no one knew very clearly how it worked. Bacon did. He shows this plainly in many passages. The next step, an obvious one, was to take two convex lenses, any two would do, arrange them properly, by theory or mere ex-

perimentation, and look into a world the very existence of which men had not suspected before.

Did Bacon do this? There is evidence that he did.
The most striking is contained in the Thirteenth
Century Voynich Manuscript, recently discovered.
The evidence that this extraordinary work was written by Bacon is not entirely conclusive, but its age has
been definitely established by a technical examination of the paper, ink, etc. Its history has been traced
back into the Sixteenth Century without evidence of
any attempt at forgery. And if it were not by Bacon,
it must have been by some contemporary of his whose
mind was even more remarkable, and whose name,
by a miracle, has perished utterly. This Voynich
Manuscript is written in a cipher so complicated that
no one was able to throw the least light upon it until Professor Newbold of the University of Pennsyl-
vania discovered that under a low-powered microscope the small letters of certain meaningless words
resolved into close-packed groups of curved lines
very like shorthand. Critics have claimed that they
were made by wrinkles in the paper, but it must
have been very unusual paper, for no such markings
have been seen on any other manuscript. Professor
Newbold spent many years trying to decipher the
strange marks, and made some progress, but died before he'd translated enough to make his solution very

convincing. However, the very existence of the marks, totally invisible to the unaided eye, proves that Bacon had at his disposal a microscope of no small power.

But the letters of the code are not the only proof. The book is illustrated with curious drawings, grotesque human figures, mystic signs, and strange shapes made up of little circles like irregular bunches of grapes. The cipher captions on the drawings are still an unsolved puzzle, but the drawings themselves are no puzzle at all. Bacon was prying into one of the deepest human mysteries, the mystery of fertilization and reproduction. He did well to conceal his work in an impenetrable cipher, for if the Church had learned what he was up to, his blasphemy would have been punished by torture or the stake.

It doesn't take a scholar or a biologist to see the meaning of the drawings. Every layman can look for himself. The little female figures are very graphic symbolization of the process of fertilization and pregnancy; the sinuous tadpole-like things are human spermatozoa, for the first time pictured by man, and the bunches of small grapes are various kinds of body tissue, complete with cellular structure and all. Bacon had looked behind one of the most sacred veils of Nature. It was lucky for him that he didn't send his results to the Pope.

So much for the scientific content of Bacon's great work. There's a great deal more, most of it of high quality and well worth looking into. But the most important part is the chapter which Bacon calls "Experimental Science." Three hundred years later, Francis Bacon, probably no relation to Roger and certainly not a descendant, wrote his *Novum Organum* propounding the principles of scientific thought. It would be interesting to know how much he got from his long-dead and almost forgotten namesake. He probably had access to John Dee's collection of Roger Bacon Manuscripts. But the problem is one which will never be solved.

The scientific method is an old story to us now. Most of us think that way, and those who do not, think hardly at all. The process seems so natural to us that we can't believe it was ever unknown. But during the Middle Ages, and indeed during most of the history of mankind, the prevailing habit of thought was very different. Men looked to the past for enlightenment. They had the "good old days complex" to an astonishing degree. "Revealed truth," "lost art," and "the great minds of the past" were magic phrases of medieval learning. The road to knowledge was through old books, the older the bet-

ter. Only rarely did men dare think that they might,
by simple observation, learn more than their predecessors had ever dreamed of.

Carefully, judiciously, stolidly, Bacon smashes
this theory of thought. He demonstrates various
errors in the greatest authorities. He ridicules the
popular superstitions which had worked their way
into the uncritical minds of the learned, and he savagely denounces the theologians of the day whose
imposing systems were built on nothing but empty
words and thin air. "Get out and observe," he urges
again and again. "Get an 'Irish crystal' and look at
the play of iridescence on its surface. See how it resembles the rainbow. There's a secret in those colors,
perhaps a secret which will do great things for mankind. Take a slate, a pencil, and a compass and see
for yourself the truth of Euclid's propositions which
young men learn to say parrot-like in the schools. Go
to foreign lands and look for the strange creatures
which are supposed to inhabit them. Experiment
with mirrors, growing plants, water, and the corro-
sive reagents of the alchemist. Above all, don't believe anything on authority while there's the
remotest chance of testing it for yourself."

This is a heavy blow at the pillars that supported
the Church and the whole medieval structure. Bacon
softens it as much as he can, giving long passages to

the usefulness of the new system of reasoning to theology. He defines science as "a perfect understanding of God through knowledge of His works." He states that his pupil John, whom he'd educated to think in the new way, was nevertheless remarkably free from sin.

But even so, Bacon doesn't feel he's given enough to Cæsar. He has an uneasy feeling that the Pope will look a few hundred years into the future, after the manner of popes, and see the Church tottering to its fall, the structure of theology discredited, and the world plunged in a vast and dangerous confusion. So he ends his book with a long and dreary chapter on moral philosophy, carefully planned to soothe the feelings of a suspicious churchman and make him see that the devil was not the motive force behind him and his theories.

In the last chapter Bacon sounds like any of the other schoolmen. He is long-winded, fatuous, and vague. He discusses theological problems which have no meaning to the modern mind. But let's forgive him this relapse. He was writing a *persuasio* for a pope, a medieval pope, the doctrinal autocrat of Europe. His freedom, his life even, hung in the balance, and perhaps he was laughing up his sleeve.

★        ★        ★

It was the year 1277, ten years after the *Opus Majus* had left so hopefully for Rome. Bacon sat alone by his furnace, the coals of which still glowed dimly and sent out a small stream of heat. He watched the fire draw down deeper into the powdery ash. No sparks flew up; no crucible shone white; no mysterious liquid bubbled in its retort, giving off with its steam the intoxicating hope of a great discovery. Bacon's diagrams were packed away; his lenses and mirrors were gone from the shelves. His books were gone too, and a great gloomy silence filled the room where once the voices of truth spoke clearly from the blackness beyond human knowledge. It was the end—the end of his hopes for Christendom and the world. The Church and Scholasticism had won; Bacon and "the True Knowledge" had lost.

Singly and silently other men came into the room and sat down on benches near the master. When all the benches were filled, they stood. Some were monks in black robes; some were friars in brown or gray. Some young men wore the clothes of ordinary laymen, while one old man in the gay cloak of a noble had drawn a corner of it over his face and was weeping silently. No one spoke; the empty shelves, the bare tables, the dying fire spoke for them.

When no more came, when silence had penetrated all corners of the room, Bacon looked up. He was an old man now, with white beard, white hair, and a bent appearance about the shoulders. He'd drawn his brown friar's robe tight around him as if he were cold. Finally he spoke.

"You've all seen the decree," he said.

Several of the listeners nodded. One young man said "Yes" softly and there was silence again.

Bacon stood up and looked carefully from face to face. He counted. Thirty-two men were in the room. Each one he knew. Each one was an old friend and trustworthy. Bacon motioned for the door to be closed, for curtains to be drawn across the windows. Then he went to a corner, pried with a knife at a crack in the wall, and took down one of the oak panels. The space behind was filled with books. Bacon took them out and piled them on a table. There were twelve in all. He turned to his friends, who still sat silent in the semi-darkness.

"You have watched the rebirth of the true learning," he said, "and now you see its burial. My efforts and yours were like a small flame held under a pile of wet faggots—too weak to light the whole, and dying out to leave only a few embers. Perhaps the embers will flame up again. I hope so. I am doing all I can to assure it. These books are the embers, and the secrets they'll teach to all who know how to read

ROGER BACON

This drawing from a fifteenth century manuscript in the Bodleian Library shows the great scientist meditating on his many troubles.

A PAGE from the famous Voynich manuscript.

them are the flames which will some day set fire to all Christendom.

"I do not know yet what my own fate will be. The general of my order has condemned me for heresy, for *'novitates suspectas.'* He has sentenced me to prison. The men-at-arms will be here in an hour or two. I may be shut up in a dungeon. I may be put in some remote monastery, kept from books, and prevented from writing. I do not know. Since the death of Clement IV my enemies have been all-powerful at Rome. An appeal would be hopeless. But before they take me away, I mean to make sure that my works shall not entirely disappear.

"My friends, please excuse my suspicion. I have suffered so much that I've learned to take precautions. There are thirty-two of you here. I have twelve books left, all my enemies couldn't find. I'm going to distribute them among you in such a way that you will not know who has received one and who has not. I trust you all, each one separately, but some of you may have a change of heart when you grow older, and there are tortures which can open the most unwilling lips."

Bacon went to the windows and adjusted the curtains so that almost no light came through. In the room could be seen only vague shapes.

"Now, my friends, come forward one by one. Do not speak. Keep your faces down. If I hold out a book, take it and hide it under your clothes. If I do not, walk on without stopping. I shall not see who you are. I shall not try to judge who is worthy of my confidence. Then if I'm put to torture, I shall be able to say honestly that I do not know who has my books."

Silently the dim shapes passed before Bacon and his pile of books. It took a long time. When all had stumbled back to their seats, Bacon drew back a curtain. The room was filled with bright sunlight. The books were gone. No suspicious bulges showed under the loose robes. Bacon stood by his empty table and looked almost cheerful.

"That was the darkest moment," he said. "Now I feel better. I can look forward with hope to what the future holds for me in the few years remaining. My friends, those books which you are concealing so loyally are my legacy to the world. Some are in plain Latin; some have their meaning slightly concealed under strange words; some are in a cipher which perhaps no one in the world will be able to unravel. Guard them carefully. Don't try to have them copied. The most skillful copyist will miss most of their meaning, for some of the letters are not what they seem.

"And I warn you, my friends, do not defend my memory in public. Carry on your studies behind closed doors, in obscure streets, with curtains over the windows and a network of twisted language to confuse anyone who tries to read your notes. I was too hopeful. I thought that the Church would see the value of my work and weigh its worth against the continuance of its own power over the mind of Europe. I was wrong. The Church prefers to keep what it has. It prefers to force Christendom back into the darkness of the past rather than to risk losing an ounce of its power.

"Some day, perhaps when we all are dead, the new learning will come to life again. It will creep into the minds of churchman and layman alike. The Pope and all his cardinals will be too weak to stand against it. And the learning of Europe will sprout like young wheat in spring."

Bacon sat down again, and drew his robe around him.

"Now go, my friends," he said, "and leave me alone. Soon they will come to take me away. I don't want them to search you for books."

When they were all gone, Bacon sat alone and in silence. He sat for a long time without motion. Then came the tread of heavy boots at the door. Bacon rose and went out into the courtyard. The footsteps died away, and the laboratory was empty. The soft voice of truth had none to listen to it now.

\* \* \*

Such was the end of Bacon's great program. What happened to him next we do not know. He disappeared from sight until 1292, when on the death of Pope Nicholas IV, his friend Raymond Gaufredi, the new general of the Franciscans, had him set free. He emerged from prison an old man and nearly forgotten. Seventy-eight was tremendously old for those days, and fifteen years of silence can erase the brightest reputation. He went at once to Oxford, where he wrote one more book. He was still firm in his heresy, and he attacked for the last time the intolerance and narrow-mindedness of the age. But his fire was gone. He had learned nothing new, and a few months later he died.

But Bacon's books were not dead. The cipher works passed from hand to hand, most of them mysteries to the present day. Even some of those in plain Latin escaped that favorite ecclesiastical festivity, the bonfire of heretical works. Hidden in monasteries where liberal abbots winked at heresy, carried about under the robes of furtive alchemists, wandering far away among the Saracens who never suppressed heretical books, they lived an underground life, but they lived.

Quotations from Bacon, generally without credit, were constantly creeping to the surface to prove that his works were still prized by the scientific-minded. *Imago Mundi* of Pierre d'Ailly has a direct, word-for-word quotation discussing the probable proximity of Spain and India, then considered the extremities of the world. This was cited in a letter from Columbus to Ferdinand and Isabella as one of the reasons for thinking that a voyage west into the Atlantic might bring valuable results.

Bacon's proposed reform of the calendar remained the best solution until Pope Gregory VIII finally made the change. Bacon's device of omitting one leap year in a century is still in use to-day as the nearest practical device. Bacon's prophecy of intellectual stagnation for Europe if his program were not adopted came true with a vengeance. The Fourteenth and Fifteenth centuries don't compare with the Thirteenth in any way. There were no men like Bacon and Grosseteste at Oxford or Paris. Not even theologians like Albertus Magnus and Thomas Aquinas. Theology had finished its work. Nothing remained to be done, and the schools of Paris declined abruptly in power while the "Doctor Subtilis," Duns Scotus, intoned lengthy sophistries and added a new word, "dunce," to the English vocabulary. The Church sank gradually lower and lower. Became a tool in the hands of the kings of France, and after the Pope returned to Rome dissolved into such worldliness and hypocrisy that it lost all spiritual and intellectual influence.

Two hundred years after Bacon's death came the great upheavals, the Reformation and the Renaissance, which went far to free both art and learning from the dead hand of the past. But until then science was driven underground. The alchemists wrote, but seldom published. They concealed their discoveries in a veil of mystical language; they crept fearfully from town to town, and carried on their experiments behind closed doors. Bacon was the last great liberal mind to speak up so that all the world could hear him.

So before we go on to the great age of free thought to come, let's take a look at those shadowy alchemists, who through a litter of kabbalistic signs and magical formulæ kept science alive.

# Alchemy

# CHAPTER III
## Alchemy

THE streets of medieval Paris were narrow and twisted, but none more so than Notary St. which led to the house of Nicolas Flamel. Near by was the back door of the Church of St. Jacques la Boucherie, but few worshipers went in that way. The upper stories of the houses beetled forward into the street until they nearly met. The paving stones were always damp. It was dark even in daytime and inky black at night. If you wanted to live in obscurity, this was the place.

The house was like most others in the walled city, dark and cramped from the painful lack of space. But if the casual visitor should penetrate by any chance to a certain inner hall, he'd notice a small door with a keyhole fully three inches long. Behind it a flight of stone stairs led down to a hidden cellar. This was the secret laboratory of the alchemist Nicolas Flamel. Sometimes through the door came the dull, pulsating roar of a furnace, and sometimes the fumes of sulphur crept out through the keyhole.

By profession Flamel was a scrivener, something between a lawyer and a copyist. He was successful, and he earned a sufficient living. But his heart was not in his work. After he'd copied his documents, fixed his seals, and drawn his contracts for the day, he'd dive into his cellar and set to work feverishly with the weird apparatus he'd been collecting for years. Sometimes he'd read long hours by candlelight, trying to extract vague directions from a matrix of misleading language. Sometimes he'd fire up his furnace, arrange his retort, and purify by distillation a liquid which might, actually *might,* contain the Essence of the Stone.

But things didn't go very well. He kept at it for years: learned Greek and Hebrew, perfected his Latin, read all the major works of the Adepts of old, penetrated deep into the mazes of the kabbala. But the more he learned the darker the mystery grew. Sometimes he'd take a friend into his confidence and ask his advice. "There's nothing in it," the friend would say, "nothing but ancient madmen trying to cover up their ignorance with difficult words." Perhaps he was right. Perhaps . . . Flamel would stay out of his cellar for a few days.

But never for long. His good resolutions to leave alchemy to others and stick to his scrivening never lasted more than a day or two. Soon he was back in the cellar working away as hard as ever. His good wife, Perenelle, came down sometimes, too, and sat by the furnace with her knitting while he wrinkled his forehead over books so obscure that their meaning lay hidden completely from his hopeful eyes, books with roots so far in the past that no one could say for sure whether men or angels had written them.

One evening he felt very low in his mind. An experiment had just failed. He'd followed directions carefully and painfully, spent weeks heating an unresponsive lump of copper with strong vinegar, spirits of wine, sugar of lead, and bone-ash. Nothing much had happened. The copper turned green, that was all. Copper usually did that. And this copper was supposed to turn red and glow with the brilliance of a ruby. Disappointedly Flamel emptied the crucible into a tub of water and sat down at his desk where an ancient book lay open. He read silently for a few minutes. Then he looked up. His wife had come down the stairs, taken her usual seat, and was working placidly at her knitting.

"It's always the same, my Perenelle," he said to his wife, who stopped work dutifully to listen. "Always the old books say they've given you full directions. Just do what they say, and the Stone will appear in your crucible. But nothing happens. Nothing but a bad smell and a lot of charcoal burned up for noth-

ing. Now listen to this. Who could follow this recipe?
Yet it's from *The Book of Ostanes,* written by the
great Aristotle himself. Listen:

> " 'A lion is reared in a forest. A man has de-
> sired to use it for a mount, and he has put a
> saddle on it and a bridle. Vainly he tries and can-
> not succeed. He is then reduced to trying a more
> clever stratagem which allows him to keep it in
> solid bonds and to put on the saddle and bridle.
> Then he conquers it with a whip with which he
> deals it grievous blows. Later he looses it from
> its bonds and makes it march like an ordinary
> creature—so completely that one would affirm
> that it had never been savage a single day.
>
> " 'The Stone is the Lion; the bonds are the
> preparations; the whip is the fire. What say you,
> O Seeker, to a description so close?'

"Close, *close!* He thinks that's close." Flamel shook
his head sadly, and Perenelle returned to her knitting. He read on for a time and then looked up
again.

"Perenelle," he said, to get his wife's attention.

"Yes."

"Here it is as clear as day, not the secret of the
Stone, but the reason I can't find it in any of my
books. I lack faith and knowledge of God. Listen to
this:

" 'The Adepts of old have defended the secret of the Stone at the point of the sword, and have abstained from giving it a name under which the crowd may know it. They have disguised it under the veil of enigmas, so that it has escaped even penetrating spirits, and so that the most lively intelligences have not been able to comprehend it, and hearts and souls have despaired of knowing its description. There are only those whose minds God has opened who have understood it, and have been able to make it known.'

"There it is. The words in these old books mean more than they seem to mean. But only a man whose mind is lit up by faith can understand them. And I have not the faith."

"You've got as much faith as anyone," said Perenelle with decision, picking up her ball of wool and preparing to go upstairs. "You just keep on reading and you'll see what it all means. Then you'll find the Stone. Others have found it. Why not you?"

After his wife had gone, Flamel put his hand to his forehead and stared at the pages before him. *The Book of Ostanes* grew dim, and the letters, so faint and old, grew blurred before his eyes. He must have faith, he thought. Faith was the thing, not knowledge or money or books. Faith, and the Stone would be his.

Perhaps he fell asleep. He never knew for sure, and it didn't make much difference, for in a spot of light which approached from the distance he saw an angel bearing a large book in his arms.

"Nicolas Flamel," said the angel, "look at this book. You wouldn't be able to read it now, and most men would never be able to. But some day you will find in its pages things which no one but you will see."

The angel disappeared in a flash of light, taking his book with him. Flamel awoke—or perhaps he was awake already, he never knew—and ran upstairs to tell the good news to Perenelle.

"I *have* faith," he cried joyfully, "for God has sent an angel to promise me a great book bound in brass. I must keep my eyes open. God sends His gifts in all sorts of strange ways. Perhaps the book will appear on the shelves of some bookseller or be brought to our very door."

For the next month Flamel kept peering into bookstores, looking for the gleam of brass among the ordinary bindings of parchment or leather. And every time a knock came on his door, he'd rush out hoping to see the messenger of God. At last a man came to his house with a large and remarkable book, bound like the angelic one in brass, with Greek letters engraved on the cover. Flamel offered two florins, and the book was handed over. Its bearer disappeared, and Flamel dived into his cellar to examine the new treasure.

It was the angelic book all right. On the title page was an inscription in letters of gold. "Abraham the Jew," it read, "Priest, Levite, Prince, Astrologer, and Philosopher, sends his greetings to the nation of the Jews long scattered by the anger of God." The pages were not of paper or parchment, but papyrus, which Flamel took to be the bark of small trees, and the letters were written in various colors. Every seventh page was a picture. The first of a virgin, the second of a serpent swallowing her up, and so on. A fine book it was, and enough to make any philosophical heart jump with joy.

After Flamel had spent an hour with his book, he decided the angel was right about the difficulty of reading it. Besides Latin there was Greek and Hebrew, not the ordinary varieties about which he knew something, but strange twisted letters, words which meant nothing, and sentences which began well enough, but melted into nonsense before they were half finished. The pictures were nice enough to look at, but they meant little to poor Flamel, who began to despair of the task he had before him.

Perenelle looked at the book too. She was very much interested, and she wasn't at all discouraged at its difficulty. It looked no worse to her than any other book.

"Go right ahead," she said, "and read it over and over again. Sometime you'll see what it means. You might have those pictures copied and show them to some of the great philosophers at the university. They know everything."

Flamel did as his wife said. He hired an artist to copy the pictures and took them to one philosopher after another. None could tell him more than he knew already—that the pictures illustrated the preparation of the Stone and were very ancient. Finally he heard of a man named Anselm who specialized in such matters. Surely Anselm could tell him if he wanted to.

Anselm could and did. He showed tremendous interest and demanded to see the book. Flamel refused to show it to him. Then he offered to explain the pictures if Flamel would give him a tenth part of the gold he made. Flamel agreed.

"The six pages of text between the pictures," said Anselm, "represent six years of digestion of the Stone. The starting point is Mercury, this God pictured with wings on his heels. He is 'fixed' and his volatility taken away by the blood of infants. Look, here is Herod slaying the children of Bethlehem. The Mercury is then joined with Sol and Luna and turned into a plant like that pictured, and afterwards corrupted into serpents. The serpents you can see plainly. And these serpents, after being perfectly dried, and digested, are made into a fine powder of gold, which is the Stone. And if you don't understand this discourse, you are not an Adept, and shouldn't be reading the book at all."

Flamel was as puzzled as ever, but pretended to understand. He went back to his laboratory and began long operations with mercury, gold, and silver. One direction he did not follow, that of "fixing" the mercury with the blood of infants. Such things were not for Christians. God had made new rules for the world since the book was written.

For years he worked on the mystery, never getting any nearer to the secret. Twenty-two years he worked, and his hair grew gray. The language of the book became clear enough to him, but the meaning no clearer than before. At the end of this time he called Perenelle for a consultation.

"I've accomplished nothing all these years," he said. "I'm beginning to think that Anselm told me nothing but foolishness, hoping to discourage me and make me give him the book. What shall I do now?"

"Ask someone else. Perhaps you'll find a learned man who isn't also a liar."

"No one in Paris knows more about this than I do."

"Then go to another place. A Jew wrote this book, and a Jew should be able to explain it. The king has driven the Jews from Paris, but there are still Jews in Spain. And they are very learned."

"You're right," said Flamel. "I shall go to Spain and find a learned Jew."

Putting his business affairs in the hands of a trusted friend, he bought a pilgrim's traveling outfit and set out for Spain. At León he met the man he was looking for, a learned Jew named Canches, who recognized the book as one of the most ancient treasures of his race and agreed to interpret the copies of the illustrations in return for a promise of a look at the book itself. Flamel made careful notes of the explanations. Then they both set out for France. On the way the Jew died, leaving Flamel alone in possession of the secret.

Now everything was clear as day. He had the fundamental secret, the *prima materia,* and the rest was easy. In three years, no more, he had samples of the Stone, both the red and the white. Calling Perenelle down into the laboratory, he made the great final experiment.

His wife watched breathlessly while he heated mercury in a crucible and threw on it a small amount of the white powder. The mercury turned to silver, bright and shining. He heated another portion of

mercury and this time threw on it the red powder, the
sacred Stone itself. There was a flash of yellow light,
and the mercury turned to fine gold—even softer,
brighter, and more pliable than the finest gold of
the mines. Hand in hand, he and Perenelle went
across the street to the Church of St. Jacques la
Boucherie and gave thanks to God for a lifetime well
spent.

Such is the story of Nicolas Flamel, and with small
variations it might tell the tale of a thousand other
alchemists from the Roman Empire to the present
day. The same features always appear—the years of
fruitless work, the angelic vision, the mysterious
book, the old magician who explains the mystery,
and finally success. The Stone is his and all the gold
he cares to make.

Naturally Flamel never told his secret. That
would be sacrilege. He'd learned it from God Himself and from long probings into the sacred shadows
of the past. To reveal it to the world would be as
wicked as to tell the secrets of the confessional.

Of course there was only one secret, the fact that
there was no secret at all. The alchemists never made
gold; they never lived forever; they never had a look
at the Philosophers' Stone. But their art did not die.

Besides being a science, it was a religion of a sort, and religions don't die from lack of success. It is one of the most touching things about the human soul that religious or semi-religious hope is not killed by a little thing like a thousand years of failure.

Anyone who looks into the subject of alchemy will be struck by the prominent position it occupied in the thought of the Middle Ages. It absorbed most of the scientific energy of the period. Among its devotees were popes and emperors, kings, bishops, and nobles, and down the social ladder to peasants chanting old charms over a glittering lump of pyrites. King Henry IV of England employed an alchemist to help him debase the coinage of Europe. Emperor Rudolph II kept a whole company of them busy at Prague. Cagliostro, who was an alchemist as well as an all-round fraud, threw France into confusion as late as the end of the Eighteenth Century.

And strange as it may seem to the scientific-minded, alchemy isn't dead yet. Public libraries have numerous calls for the same books which mystified Flamel. In every modern city live men who puzzle over the same old enigmas, try the same old experiments, with modern apparatus perhaps, but following the same old principles which are older by far than modern civilization. In 1929, in the most scientific country on earth, an alchemist got large sums from General Ludendorf by promising to make enough gold to restore the Kaiser to the throne of Germany.

Alchemy was the father of chemistry, and it's hardly surprising that the father continued to live in fair health long after the birth of his eldest son. It's hardly surprising either that the son, chemistry, showed in his early years a good many resemblances to his father. Without some slight knowledge of alchemy and the intellectual soil in which it grew, it would be impossible to understand such men as Paracelsus, Boyle, and Priestley, all great men, all ranking high in the history of legitimate science, and all showing the influence of alchemy.

The story of alchemy takes us far back into the remote past. There is good evidence that it came first, like so many other things, from China. Certainly the Chinese were hard at it long before it was heard of in Europe. Chinese alchemy was an outgrowth of Taoism, which was founded in the Sixth Century B.C. by Lao Tzu, the "Venerable Philosopher" who was born after a gestation of eighty years with a long white beard and the features of an elderly man. The vague and peculiar theories of Lao Tzu do not concern us here, but after a short period his followers developed a mystical science whose resemblances to European alchemy are so striking that they need little pointing out. Both, for instance, had two main objectives, to make gold from base metals and to prolong human life. Both believed that metals "grew" in the earth. Both used the same mystical expressions such as "philosophical" mercury and "philosophical" lead. Both searched for a Philosophers' Stone. And Chinese and Europeans alike clothed their work in a figurative language designed to keep its real meaning from the vulgar. Or, as the vulgar said, to conceal the fact that there was nothing to conceal.

The date of alchemy's arrival in the West confirms strongly the theory of its Chinese origin. There had been a good deal of "gold making" in Egypt from the earliest times, many recipes of which have survived, but Egyptian "gold making" was not true alchemy, only a process of adulteration still practised no doubt wherever cheap jewelry is manufactured. True alchemy didn't stoop to such practices, at least not in theory, and true alchemy didn't appear in Europe until an embassy from Marcus Aurelius Antoninus (An Tun, the Chinese called him) arrived in Indo China in 166 A. D. and established trade between the two empires. Alchemy moved west with the silk and jade of China, appearing first in those cities, Alexandria and Constantinople, which controlled the trade with the East. By the end of the Third Century it was so popular throughout the

Roman world that the Emperor Diocletian, fearing the effects on the currency if the alchemists should succeed in making gold, ordered all books relating to it to be destroyed. This prohibition, like most others, had only a temporary effect, and alchemy, becoming more mystical as it grew in popularity, took its place with the numerous Eastern sects which struggled shrilly with Christianity for possession of decadent Rome.

The history of alchemy is one of the most baffling and maddening subjects in existence, and each scholar who is rash enough to grapple with it emerges with a different conclusion, or more frequently none at all. Some consider it a religion and work it into the gaudy fabric of theosophy and the "secret doctrine." Others treat it as an embryo science and try without success to translate its mystical language, its Green Lion, its Goose of Hermogenes, into sober chemical formulæ. Others call it a "popular delusion" and let it go at that, which of course is begging the question. Alchemy was too important, has played too large a part in the intellectual life of a thousand years, to be so lightly put aside.

Most modern scientists look down on alchemy from a great height. Scientists are notoriously narrow-minded in some respects, and their attitude toward any system of thought except their own is usually intolerant. The alchemists, judged by modern standards, look very ignorant and foolish, but we have no right to judge one age by the standards of another. Alchemy, with its mysticism, its numerous frauds, and its habit of degenerating into a singularly unsatisfactory kind of religion, was the product of two tremendous handicaps with which modern science does not have to contend. These were an unproductive and backward-looking intellectual philosophy and the unfortunate choice of objectives impossible to attain. This combination would be enough to discourage any scientist and turn him into a mystic.

Alchemy didn't get started until the classical age was long past its intellectual prime. Decay had begun, and while the barbarians of the West were battering the frontiers, the religions of the East had already invaded and conquered the intellectual empire of Greece and Rome. And wherever they touched they cast a blight from which the world did not recover for a thousand years or more. Christianity wasn't the first of these or the last. It was merely the winner. Isis led the ancient cults of Egypt to the glorious assault. Mithra stalked out of Asia. The Greek mysteries, long kept down by the deadly weapon of tolerant amusement, began to be taken seriously. Judaism, harsh and cruel as always, gained much power if few converts. Even the philosophy of the Greeks slid backwards into the silly confusion of almost-Christian Neoplatonism. Aristotle, no doubt, turned in his grave daily.

By the time the barbarians actually arrived to overwhelm the Western Empire, the classical culture had developed the habit of looking backwards for its inspiration and authority. This was natural and inevitable. For a long time the world had been on the down grade and everybody knew it. Civil order had disappeared; anarchy was spreading. Language and literature were sinking. The military power of the empire had evaporated. And the tolerant old religion which pleased the common people without hampering the educated was dying under the assault of various cults which neither believed in toleration nor practised it. The poets tried feebly to imitate Virgil; the philosophers, such as they were, repeated dimly the thoughts of the old masters; the jurists simplified the law, threw away its finer subtleties, incorporated barbarous doctrines better suited to barbarous minds, and watched the highly organized Roman Empire sink to a confusion of wandering tribes and isolated villages. No wonder men looked backward. In the present was confusion; the future promised worse. Only in the past were order, peace, beauty, and knowledge.

Such was the background of the Roman Church, the mighty institution which arose to organize Europe on a new pattern. It looked backward to a golden age in the past, and forward only to glory in another world. "No use to teach hope for this world," said the Church. "This world is past hope. Things are getting worse and will continue to do so. Soon will come the final catastrophe. Gabriel will blow his horn, and the whole futile business of worldly life will pass forever into the hands of a supernatural receiver. The next world will be better for some and worse for others. But there's no hope whatever this side of the grave."

Gabriel, of course, didn't blow his horn on schedule, and as the years passed, people lost faith in the impending destruction of the world. But they didn't stop looking backward. The habit was too strong. They could see for themselves the dim outlines of great things gone forever. Their own efforts were so feeble in comparison. They pored over the surviving fragments of ancient learning with worshipful veneration. Perhaps if they searched, and studied, and tried to understand, they might approach if not equal the learning of the great men who'd been dead for so many years. The Church tried to focus this attention on the Scriptures and the Church Fathers. To a great extent she succeeded. But these authorities are notoriously deficient in practical information, and other books existed, some of them far older than the Church or its Fathers. To these turned the groping medieval scientists. Medieval Alchemy was born with its eyes in the back of its head. It hoped to discover the secrets of the past—nothing more. No research as we know it. That would be useless. No hope of progressing beyond the discoveries of the past. That would be folly and presumption. If they could only regain what had been lost, that would be enough.

But this timidity, this lack of faith in their own ability to discover new facts, was not alchemy's only handicap. There was still another and a greater—the obsessed determination to make gold. The history of this attempt carries us back to the first stirrings of civilization.

Gold was always admired and coveted. It was the only available metal which remained bright in use. Its color identified it with the sun, which all primitive races worshiped in some form. It was easily worked into ornaments, and it became the chief medium of exchange almost as soon as money was invented. Greed, love of beauty, religion, and political economy combined forces to make gold the most desirable of primitive commodities.

There was no real reason to suppose that gold could not be made from other materials. Other metals could. Bronze, for instance, which differs from copper and tin as much as they differ from each other. And useful substances like glass which existed nowhere on earth until men learned to mix certain earths and melt them together. The modern critic protests that these substances are not elements like gold, but the idea of a chemical element appeared only in recent times and grew very slowly to its present authority. The reasoning of the ancients was perfectly logical. They made various things they wanted. Why couldn't they make gold, which they wanted most of all?

So the alchemists set to work—with a supply of hope and determination which lasted several thousand years and is not entirely gone to-day. At first their methods were simple. They experimented with alloys, hoping to find that gold was a mixture of other metals. Certain alloys, such as those containing copper and antimony, resemble gold very closely. But none was satisfactory, and the alchemists moved on to other fields.

They gathered rare minerals and tried to extract gold from them, reasoning correctly enough that if copper exists in the earth both as a metal and as an ore, gold might do the same. No luck here either. The minerals yielded various strange things, but no gold.

So the alchemists arranged the known metals according to their resemblance to gold. There were "base" metals and "noble" metals. Lead came lowest. It tarnished almost at once and had little brilliance, malleability, or resistance to superficial oxidation. Next came iron. It tarnished, too, but was more brilliant and malleable. Then copper, and so on. Silver was next to gold, almost as noble. And gold was the king of metals, perfect, shining, incorruptible. It seemed likely that if Nature had made the metals in a series leading gradually up to gold, she must have some way of changing one metal into another.

Here's where the doctrine of the Philosophers' Stone entered to complicate the situation. The Stone was the marvelous reagent which could "transmute" a base metal into a noble one. The idea is a very old one, and it became very early the central doctrine of alchemy. The secret was supposed to have been lost centuries ago, or to be known only to certain "Adepts" who were jealously keeping it from the public. The Stone might be anything on earth, or something not on the earth of ordinary human experience.

First the alchemists looked among the minerals. They tried every conceivable reagent which could be made from them. They had great faith in long-continued heat and developed furnaces which could maintain a high temperature indefinitely. Sometimes they heated the same mixture for years on end, although this meant keeping a man at the bellows day and night. Still no result.

Far from being discouraged the alchemists started off on another tack. Nature is one, they reasoned. She is a unit, indivisible. What she does with living plants she can also do with the metals which grow in the earth as plants grow on the surface. Therefore the Philosophers' Stone may be some sort of vegetable substance which speeds up the growth of a metal as manure speeds up the growth of a plant. They searched high and low for the "philosophical manure," tried every conceivable vegetable substance they could lay their hands on. Still no gold.

Next they tried animal substances, for animals contain even more of the mysterious essence of life than plants do. They collected dried toads, small black lizards, blood of bats, human brains, eyes of the basilisk from the deserts of Africa. They tried all these things, still full of hope, still sure that the secret lay only a little beyond their reach. But they got no gold.

By this time a good many of the alchemists had given up purely material methods and branched off into magic and mysticism. The Church taught that

God had worked miracles in the past and could do so still if He chose. He certainly could give the secret of the Stone to those who loved and served Him. So the alchemists became ascetics, monks, and hermits. They fasted rigorously and divided their time between prayers and experiments, hoping daily that God would modify some simple reaction and make it produce the Stone instead of ordinary cinnabar or verdigris. But God paid no attention. The strictest hermit got no more gold than the merriest worldling who worked in the tavern and hired the pretty barmaid to tend his furnace.

"Well," said the alchemists, "perhaps the Church has been deceiving us all these years. Perhaps we've been praying to the wrong God. There were faiths before Christ, before Moses, before Abraham. Perhaps the old gods can do what the God of Holy Church cannot or will not. Let's give the old gods a chance."

With faltering steps and many misgivings they dug into the darkness behind them and resurrected the long-dead gods of the past: Zeus, Moloch, Baal, Isis, and Indra. Secretly the old books stole out of their hiding places; old rites were celebrated, and the alchemists fearfully, in terror of excommunication or worse, set up hidden altars to strange deities. But Moloch helped them no more than Jehovah, and

Isis produced no more gold than the gentle Jesus, who probably didn't approve of alchemy anyway.

The next step was a terrible one, and a man had to be brave to take it. The Church taught that God was all-powerful—but was He? There was evil in the world, plenty of it. This must mean that the devil was still active in spite of the power of God. Perhaps the devil, not God, had jurisdiction over the metals in the earth. It was worth trying anyway. There were plenty of devils. Holy Church herself taught that, and it ought to be easy to get their attention.

Black magic this was, and dangerous. The alchemists worked secretly in caves, in the depths of forests, in cellars and abandoned dungeons. They wrote charms in virgins' blood, recited the Lord's Prayer backwards, celebrated the Black Mass and the Witches' Sabbath. But still no gold. Neither Satan nor his lesser devils appeared to help them.

This was the end. There was nothing more to try, and genuine alchemy died gradually of discouragement. The mystical side of the art lived on and still lives to-day in the doctrines of theosophy and other cults. The alchemist charlatans continued to fool the gullible with stories of sudden wealth gained by following the directions of some mysterious book. But the traditional alchemy declined from loss of hope.

A TYPICAL ALCHEMICAL RECIPE

My Son, I was dead without thee
And lived in great danger of my life.
I revive at thy return
And it fills my breast with Joy.
But when the Son entered the Father's house,
The Father took him to his heart
And swallowed him out of excessive Joy.
And that with his own mouth.
The great exertion makes him sweat.

From the "BOOK OF LAMBTREE"

Its devotees drifted into other sciences. And when
Paracelsus burst out of Switzerland shouting that
the true object of alchemy was not to make gold, but
to prepare medicines to cure the ills of mankind, the
best of the alchemists followed him in the first attempt to organize into a real science the rich accumulation of chemical facts which the alchemists
had discovered in the course of their search for gold.
A thousand years they had worked, two thousand perhaps, without material pay, without the faintest
shadow of success. For a thousand years they had
served a cruel mistress who never gave them a single
smile for all their devotion. Fools, most people call
them now, and fools they were perhaps. But if they
had worked less devotedly, less hopefully, the world
would be without the science of chemistry which
means so much to it now.

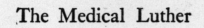

The Medical Luther

# CHAPTER IV
## The Medical Luther

IN THE Sixteenth Century all the southgoing traffic of Germany headed for the Brenner Pass, for at the Brenner the Alps are absent-minded, leaving an easy passage between northern Europe and the rich plains of Venetia. It was an age of great coming and going and the Brenner Road hummed with activity. In Italy the Renaissance was at its apex, spreading new ideas, new riches, new luxury over the gloomy North. In Germany the Reformation was getting under way with its train of war and confusion. The Brenner connected these two centers of change, and the concentrated life of Europe crowded up the narrow valley, over the low pass, and down to the plains on the other side. Troops of soldiers, trains of pack horses, papal legates, merchants, gypsies, wanderers of all sorts—all came through the Brenner. The towns along the road were the grandstands of Europe. From their doorways could be watched the whole glittering pageant of a reawakened world.

At Innsbruck the road branches, part going west to Augsburg and the Rhine, part continuing down the valley of the Inn to the Danube, Vienna, and Prague. On this lower branch stands the little town of Schwartz, on one side a fertile valley, on the other the mountains crowding out into the river as if they resented its passage through them. It's a sleepy little place in this Twentieth Century, although the Brenner, steel-shod now, still carries the freight of Europe. But in the early Fifteen Hundreds it hummed with feverish activity. Veins of ore had been discovered in the surrounding mountains, and the counts of Fügen were exploiting them with all the uneasy vigor of the age. The forests were full of charcoal burners, the valley was blue with wood smoke, and the travelers on the great road stopped curiously to peer at the glowing forges and the copper bars piled ready for shipment down into rich Italy below the pass.

Along the dusty road from Innsbruck one summer day in 1516 came a foot-traveler with a battered knapsack on his back. He walked with an eager birdlike gait, head moving a little with each stride. He was slender and sickly looking, perhaps twenty-two years old, with the thinnest sort of scraggly beard growing in small patches on his chin. His clothes were rough and coarse; he wore no hat, and his head

projected forward as if his mind were more anxious
than his body to reach its destination. He looked ordinary enough in the distance, but when he came
near it didn't take much penetration to see that this
was no common wanderer. From his thin face the eyes
looked out with such devouring eagerness that travelers who passed him on the road looked back and
stared as if they'd seen a ghost. He was Theophrastus Bombast—Phillipus Aureolus Theophrastus
Bombastus von Hohenheim, surnamed Paracelsus,
to use his full name, which he never did—son of a
doctor at Villach in Carinthia, student of alchemy,
medicine, and other things, and collector of miscellaneous knowledge along the by-ways of Europe. He
stopped in the town square, inquired for the manager
of the mines, took a folded letter from his pocket,
and disappeared into one of the smelter buildings.
The mine school of the Fügers had gained its most
famous pupil.

The outfit at Schwartz would now be called an "industrial research laboratory." It was not the first of
its kind by any means. The Fuggers of Augsburg,
for instance, had maintained a mine school at Villach for years, and Paracelsus's father was the official
doctor who treated burns from molten metal and
crushed hands from falling rock. But Schwartz had
a feature which made it almost modern. In a build-

ing just outside the mine mouth worked a group of theoretical alchemists, distilling, subliming, and calcining to their hearts' content. The counts of Fügen were practical men with little hope of finding the Philosophers' Stone, but they did hope that the alchemists would extract copper, zinc, or something else salable from the promising but refractory ores which veined the mountainside. And in holding this hope, they were doing just what the General Electric and the Du Ponts do to-day. Paracelsus found at Schwartz the meeting place of theoretical alchemy and practical chemistry. It was this opportunity which started him off on his career of injecting new and somewhat hectic life into the science of his day.

He was a strange character, this wandering student. Like so many great men, he was one of those freaks whose eccentricity keeps them from sinking down into the soft featherbed of comfortable mediocrity, like a crow with a white wing feather unable to live in peace with his normal fellows. Paracelsus was restless, unfriendly, a life-long wanderer, almost an outcast. His friendships were all on the intellectual plane, never on the normal plane of ordinary human affection. He was foul-mouthed, vindictive, utterly tactless, leaving behind him a trail of enemies buzzing like angry bees and swearing murder and sudden death if he should ever return.

There's a story which would account for his extraordinary bitterness and hatred for all the world. It sounds unlikely on its face, but is so persistent and so backed up with indirect evidence that it may be true.

The tale goes that when Paracelsus was two years old, the backwash of a small war came through the village of Einsiedeln where his father was the local doctor. Soldiers were quartered in the houses of the villagers, and the house of the doctor was afflicted with more than the usual quota. The whole troop made it their headquarters during their stay and proceeded to enjoy themselves. They got drunk, canvassed the town for attractive young women, broke the furniture, maltreated the servants, and were generally true to type. The night before they moved on, they had a grand party in the doctor's house. There was no feminine entertainment. The townspeople had fled to the shepherds' huts in the mountains. Only the doctor stayed behind, mindful of his duty to the sick and hoping the soldiers would do no harm to an old man and his fatherless boy. He went to the child's bedroom and waited in terror while the party grew drunker and noisier as the night wore on.

Toward midnight things got a bit dull. There was

nothing more to drink. The soldiers hunted around the house for amusement, and one of them came on the doctor's case of surgical instruments. He took them out and tried them on the woodwork. Then he had a drunken inspiration.

"I'm a surgeon myself," he cried. "I used to work for a barber. Come here, you men, I'll bleed you."

None of the other soldiers would let him experiment on them, but the idea was too good to waste. They rushed up the stairs to the bedroom, knocked the old man unconscious, snatched up the child, stretched him out on the dining-room table and castrated him with one of his father's scalpels. Then, frightened at what they'd done and somewhat sobered, they ran out into the street. The father came down just in time to save his child from bleeding to death, nursed him back to health, and watched him grow up into a sickly young man condemned to a life of abnormality and solitude.

Such a story seems hard to believe, but in view of the general character of professional soldiers and the general character of the Fifteenth Century it is far from incredible. And there are many things about Paracelsus to indicate that he was abnormal sexually, whether because of violence or from natural causes. He never had a real beard although some of his portraits show a feeble attempt at one. His skull, which is still preserved, resembles a woman's more than it does a man's. He never married. He hated women of all sorts. And although his dissoluteness was shouted by his enemies from the housetops of Europe, he was never accused of lechery, the favorite vice of the Renaissance.

A man with such a handicap lives a little outside the world. The ordinary roads to happiness and contentment arc closed, and he's apt to throw himself into some abstract activity with all the strength that's in him. Not a pleasant character, this Paracelsus. Not a man for cheerful conversations by the hearth. Not a man to drink beer with on warm summer evenings. But just the explosive needed in the Sixteenth Century to blast alchemy, chemistry, and medicine out of their destinationless ruts.

The two groups at Schwartz, alchemists and mine workers, didn't think much of each other, and Paracelsus didn't think much of either. "Gold cookers" he called the first, "blacksmiths" he called the second. He worked alone, taking no one into his confidence and almost rejoicing in the hatred stirred up by his bitter tactlessness.

In general he took more interest in the practical work of the metallurgists than he did in the vague fumblings of the alchemists, for a very brief examination told him that the "gold cookers" were accomplishing nothing at all. He'd rush into a mine when the news came that a new kind of ore had been discovered, secure a sample, take it back to his laboratory, crush it in a mortar, and try all sorts of reagents on it to see what would happen. Needless to say, neither he nor anyone else at the time had a clear idea of the nature of chemical composition, but certain simple observations could be made. Ores differed widely even when they contained the same metal, and the mine workers treated them in different ways. Some they roasted in the open air before smelting in the furnaces. Some they did not. Paracelsus tried hard to discover the reasons, but he got little information. The metallurgists followed traditional rules developed by the trial-and-error method. They hadn't the faintest idea why their operations failed or succeeded.

"Now Herr Bleiberg," he'd say to a foreman standing beside a pile of green ore, "what kind of stone is that?"

"That's green-stone."

"What metal comes from it?"

"Copper."

"How do you get this copper?"

"We roast it in the open air for half a day; then melt it with charcoal in the tall clay furnace."

"But there are other green stones."

"Yes. Some we treat the same; some differently."

"How do you tell them apart?"

"Sometimes by the weight. Sometimes we can't tell at all."

"Then what do you do?"

"Try everything and hope something works."

"And does everything fail sometimes?"

"Often."

"Do the alchemists help you any?"

The foreman would laugh derisively. "Those old fools! They heat the stuff for a month, make a lot of smoke, and get a mess they call 'philosophical copper' that a tinsmith wouldn't take for a gift. To hell with them!"

In a long low building worked the alchemists. Numerous furnaces stood about. The air was thick with fumes and sour with suspicion. When Paracelsus entered he would be met with a wave of hostility.

"Well, cooks," he'd sneer, "how's the soup to-day? Any gold yet?"

The outraged alchemists would turn on him angrily. "Throw him out," would come from the back of the room.

"Let me by," Paracelsus would cry. "I've got as much right here as you have. My furnace is here, and I've got lots of stuff you could use if you knew how. I haven't any old books to stick my nose into all day long, and you'd be better off if you hadn't either. Now let me by. I've got work to do."

He'd go to his corner and start grinding up ore with his mortar and pestle. At intervals he'd grin over his shoulder.

"Look, Brother Heinrich, there's gold in your crucible. Philosophical gold, too, worth a thousand ducats an ounce."

But Heinrich had been fooled too many times. He would glare into his furnace, swallow his rage, and hope for revenge. Such a scoffer! No wonder the great secret eluded them. Half the work was developing the proper faith, the proper reverence for the Adepts of old. You couldn't develop faith with that gadfly buzzing around.

Paracelsus worked hard, and the rocks of the mountainside rewarded him well. He began to classify them by their reactions. This one gave off an "air" when stirred up with oil of vitriol. This one turned black in the furnace. This one gave bright crystals of copper when dissolved in *aqua fortis* and stirred with an iron rod. No real discoveries yet. No light on the deep mystery of why the ores changed in the furnace into metal and slag. The subject was too vast. He realized he'd only begun.

Day by day the alchemists grew more hostile.

Paracelsus did nothing to conciliate them. Perhaps he didn't know how. Scornful as ever, he peeked over their shoulders, laughed at their prayers and incantations, mocked the mystical language of their books. One morning the crisis came. They moved on him in a body, grabbed him by the collar and trousers, and pitched him out into the sunlight. He rubbed his bruises, shouted back some elaborate curses, and walked off resentfully. Perhaps he ought to complain to the manager. Keep up his prestige. But what was the use? What did he hope to discover in that stinking place? Nothing but a few practical rules to enrich an ignorant count who'd spend the money on his soldiers or his mistresses. Bah! It wasn't worth doing. No inspiration in it. No reward for the spirit. He sat down on a pile of slag and looked into the valley where the Brenner Road lay shimmering in the heat, dusty with the coming and going of Europe.

As he sat and watched the traffic below, a great eagerness came over him to be on the road himself —to Bavaria, Bohemia, the plains of the North, perhaps the distant countries beyond. "There's my university," he thought. "There are my teachers. They know a lot, and they won't tell me lies."

A tinsmith came by on a donkey hung about with pots and pans. Paracelsus looked him over carefully.

"That man now! He's old and poor and driven from place to place with blows and curses. But he knows about solder and fluxes. He knows what you can do with copper and what you can do with lead. He can teach me."

Two merchants came next with pack mules heavily loaded. A soldier rode ahead and one behind. Paracelsus watched them into the distance.

"From Venice, probably. Valuable goods, too, or they wouldn't have the soldiers. Silk and brocade from Cathay. Spice from the Indies. Merchants see a lot. They know the Turks and the Tartars, the Persians and the Hindus. They've seen the holy men on the banks of the Holy Ganges. They can teach me."

Then a troop of cavalry riding up to the pass and Italy. Glitter of armor, clatter of swords, froth on the horses' mouths, and a cloud of dust trailing off down the road. Paracelsus shuddered. Those savages, scattering murder and rape—but they, too, know a lot.

"They know wounds and death," he thought. "They've seen the moment when the soul flies out of the body. They know which thrusts kill, which wounds heal and which fester. They can teach me."

He sprang to his feet and ran to his lodging to gather up his few belongings—knapsack, dagger,

heavy boots, a case of drugs and surgical instruments. Then he went to the alchemists' laboratory
and stuck his head in at the door.

"Good-bye, you pastry cooks," he shouted. "You're
rid of me now. Good-bye, you swindlers. I'm leaving.
Good-bye, you stinking sulphur burners. I'm going
to learn from gypsies, robbers, hangmen. They all
know more than you do. Good-bye, you blind moles.
The mines will be my books and the taverns my
university. Boil your gold powder. Raise your
stenches. I'll be gone."

A shout of rage came from the laboratory, and
the alchemists rushed out, blinking their eyes like
owls in the sunlight. But Paracelsus was on the
Brenner Road, his feet shuffling the dust, and his
thoughts reaching out ahead to Germany, to Sweden,
to all the world.

The road taught much in those days. Theory
among the learned was weak, moribund, paralyzed
by mystical fairy tales and the search for a non-
existent Philosophers' Stone. But as in the time of
Roger Bacon, the working people possessed practical
knowledge of tremendous value. In their daily work
they used principles of chemistry, physics, and
medicine which the serious scholars in the universities had never heard of. Glass workers knew secrets
of chemistry, midwives secrets of medicine, and

gypsies knew secrets of hypnotism and dark secrets of what would now be called "abnormal psychology." Even at the present time there's plenty of such knowledge still to be gathered, but in the Sixteenth Century the surface had hardly been scratched.

Up and down Europe went Paracelsus, paying his way by healing the sick, giving chemical information to those who could use it, and selling drugs of his own preparation. He had a natural talent for medicine, and although without formal training, was already a skilled physician. He had, moreover, an asset possessed by few doctors of the period—the habit of accurate and searching observation. "The books of the doctor," he'd say, "are the sick," and he followed his own advice rigorously. Nothing got by him. He'd ask all sorts of questions which the other doctors considered silly or useless. What the sick man dreamed about; whether he was married; what he thought about religion; what he did for a living; what his mother and father died of; whether he had any children; whether he drank wine or beer. Nothing was too small or remote for the penetrating curiosity of Paracelsus. He observed each case minutely and filed the experience away in the back of his mind to use when the chance came. Every tavern, every village had at least one sick person, and each morning, setting out on his road, he was richer in practical knowledge and better able to live up to the reputation which began to grow around him.

In every good-sized town he came to he was sure to find after a little discreet inquiry a group of alchemists working vaguely but hopefully on the age-old problem of the transmutation of metals. There were fashions in procedure, but the spirit of the science had changed little through the centuries. The oldest authorities were still considered the best. Hermes Trismegistus and the *Turba Philosophorum* were thumbed over reverently as the most sacred revelations of the art. Various interpretations were given to murky passages, and most of the passages were murky. There were several contentious schools of thought. But one thing all the alchemists had in common. They were all uniformly unsuccessful. No gold appeared in their crucibles; no sacred Stone gleamed like a ruby among the cooling ashes of their furnaces. They had worked for a thousand years, perhaps two thousand. But no slightest success had rewarded their efforts.

Paracelsus ingratiated himself with the various groups, learned what he could of their secrets, and fared forth again on his travels. Each time he became more convinced that alchemy was dying of discouragement and under-nutrition. It needed a new

symptoms and effects of disease, they preferred, with the lack of self-confidence so characteristic of the period, to dig back into the past and collect such fragments of Greek medical science as had been saved by the Arabs and transmitted to Europe through Spain. Galen was the most authoritative of the ancient writers and his hundred treatises with their Arabian commentaries formed the framework of the medieval theory of medicine.

The principles of Galenism were simple enough. The body was supposed to contain four fluids or "humors" analogous to the four elements of Aristotle—earth, air, fire, and water. Disease was caused by an excess or deficiency of one of these. The cure, usually consisting of bleeding, purges, baths, and decoctions of aromatic herbs, aimed to restore the balance. This is wrong enough to start with, but by the Sixteenth Century Galenism had accumulated a mass of superstition which almost concealed the original theory. Every religion in the known world contributed charms and exorcisms. Every plant that grew had a place in the pharmacopœia. A numerous hierarchy of assorted devils presided over each disease. And the doctor who tried to apply a simple and rational treatment was denounced as an ignoramus who didn't understand the fine points of his trade.

Such was medical science in the youth of Paracelsus—a backward-looking dogmatism, far from reality and getting farther all the time. From the Orient, from newly discovered America came strange diseases. Syphilis swept across Europe like a plague, twisting the bones of king and peasant alike. The rapid economic changes of the period brought new ailments to light. Lead poisoning from the increased use of paint, gangrene from the ragged wounds of gunpowder warfare. The Galenic doctors had no new remedies. They turned over the pages of Avicenna, Mesue, Averrhoes, chanted the age-old formulæ, prescribed the age-old remedies, and watched their patients die with discouraging regularity. Here and there a doctor would stumble on a new drug or treatment which gave good results. But like as not he'd look through the books of the ancient authorities and, finding no precedent, cast it aside for fear of being denounced by the orthodox.

Imagine a large warehouse full of goods of all sorts jumbled together in small rooms, uncounted, unclassified, unknown to the public. And a department store with plenty of customers, plenty of money in the till, but selling nothing but dirty old rags, broken bottles, and generally failing to deliver even these to the customers who paid for them. Such were alchemy and medicine in the Sixteenth Century.

The alchemists knew a great deal about what we now call chemistry, more than they're generally given credit for, but they made no practical use of their facts. The search for the Stone was a will-o'-the-wisp leading them deeper and deeper into a quagmire of profitless mysticism. Practical knowledge of chemical compounds was regarded as merely a means to an end, and an end which had never been even in sight. Medicine on the other hand was intellectually moribund. The doctors did no research at all, relying on the remote past for their information. The whole science was at a standstill. The doctors followed the traditional procedure, shut their eyes to new discoveries, and repelled with righteous indignation any suggestion that they learn from the old wives, traveling quacks, and gypsies who often cured when they failed. The doctors needed a new stock of knowledge, new drugs, new methods. The alchemists possessed just these things. Paracelsus, traveling around Europe with his eyes open, was the first to break down the rigid wall which separated the two sciences.

He didn't stay long in one country, but wandered about as he heard of new diseases, new cures.

"Sicknesses [he said] travel here and there through the whole length of the world and do not remain in one place. If a man wishes to un-

derstand them, he must travel too. Does not
travel give more understanding than sitting
behind the stove? A doctor must be an alchemist.
He must see the mother earth where the minerals grow, and as the mountains won't come
to him, he must go to the mountains. Is it a
reproach that I have sought the minerals and
found their mind and heart and kept the knowledge of them fast, so as to know how to separate
the clean from the ore? To do this I have come
through many hardships."

He didn't avoid the universities, but treated them
as sources of information, less reliable perhaps than
the simple people of the road, but not to be despised.
He never stayed at one very long, draining it dry
of information and stirring up such hatred with his
unconcealed contempt that he soon had to be on
his way again.

"Universities do not teach all things, so a
doctor must seek out old wives, sorcerers, wandering tribes, old robbers, and such outlaws and
take lessons from them. We must seek for ourselves, travel through the countries and experience much, and when we have experienced
all sorts of things, we must hold fast to that
which is good."

Such a theory takes a man far afield. He went to
Sweden to pry into the secrets which made Swedish

iron the best in Europe. To the Cossacks who smear
their wounds with a mixture of gunpowder, pitch,
and brandy. To the Tartars who make healing
medicine out of mares' milk and herbs. He went as
a military surgeon with the Venetian fleet to the
distant shores of the Mediterranean. He journeyed
with a Tartar prince from Moscow to Constantinople. He lived several months with a Turkish
magician. No country was too distant or dangerous
for the restless curiosity of Paracelsus.

All this time his reputation was growing. His
fame as a doctor traveled ahead of him. The sick
gathered around him, and he worked miraculous
cures, much to the annoyance of the local practition-
ers who generally made it so hot for him that he
could stay in a place no longer than a few weeks.
"I pleased no one," he said, "but the sick whom I
cured." The doctors spread lies about him, the
apothecaries refused to sell him supplies, and the
university authorities cross-questioned him for proof
that he practised black magic. Horrified physicians
read passages from Galen to show how the heretic
doctor diverged from established practice. Often
the opposition took violent forms. He fled from
Poland, from Lithuania, from Prussia, a mob of the
outraged orthodox at his heels.

About 1525, when Paracelsus was thirty-three

years old, he began to want to settle down, practice medicine in peace, and digest what he'd learned on his travels. So he looked around for a place which might receive him, and chose Tübingen in Württemberg, where he gathered pupils from among the students at the university. Soon he was one of the most popular teachers of medicine. His doctrines began to gain ground, and he thought he'd found a safe harbor at last.

But it was the same old story. He made no attempts to conciliate the orthodox. He scoffed at their antiquated methods, and healed the sick whom they'd given up as hopeless. After two months he was driven out with violence. He went to Freiburg. His pupils followed, glowing with enthusiasm for the new science he taught. Again he had to leave. He went to Strassburg where a university was being founded. No luck there either. Although he bought a citizenship and became a member of the surgeons' guild, he found the forces of orthodoxy arrayed against him and had to fly before threats of arrest and imprisonment.

Sitting gloomily beside the fire of a wayside tavern a few days later, Paracelsus thought things over and decided he'd have to change his tactics. All his life he'd played a lone hand, asking no help and receiving none. And the forces of medical and scientific

orthodoxy were far too strong for one man to defy
safely. The doctrines of Galen and Avicenna were
as firmly established as the doctrines of the Church
herself, as immutable, as blasphemous to criticize.

"What the world needs," thought Paracelsus, "is
a medical Luther." Then quickly came the corollary
thought: *"I am the man.* I've learned from Nature
as Luther learned from the Bible. I've burned my
Galen as Luther burned the canon law. I've only to
post my ninety-five theses on the church door."

All during the lifetime of Paracelsus, the storm
of the Reformation had been gathering force. He'd
met with it here and there—small battles between
Catholics and heretics, villages in flames, heaps of
white ashes in town squares. He'd remained neutral
as much as possible and carefully avoided centers
of conflict. But the confusion grew, and soon all Germany was involved. The peasants of Swabia and
Franconia rose in a body, pulling down noble and
prelate alike, and chanting strange half-pagan rites
of their own invention. They were defeated and savagely punished, but the conflict grew and spread.
Every town had its warring factions. Princes took
sides. Papal legates thundered from strongholds of
Catholicism. Bulls of excommunication flooded up
from below the Alps. But every day the revolt against
the old orthodoxy grew in power. Paracelsus con-

sidered carefully and threw in his lot with the Lutherans.

In Switzerland the Reformation had taken strong hold. Its leader Zwingli was second only to Luther in power. His followers penetrated to every city and canton, carrying with them new doctrines, new ceremonies, new ideas. At Basel, just over the border from Germany, the reformers had the upper hand, controlling the municipal council which governed this free city. So to Basel came the famous Erasmus of Rotterdam, the great humanist whose writings had done so much to prepare the ground for the Reformation, to spend his old age with his friend Johann Froben, publisher and bookseller, and humanist almost as famous as himself. From Basel, one of the few safe spots in Europe, Erasmus watched sadly while the conflict started by his ideas grew into a savage struggle which reduced to near-anarchy the peaceful, tolerant Europe of his dreams.

For some time Johann Froben had been suffering from a slight injury to his foot. It wasn't much to start with and would probably have cured itself. But the old-fashioned doctors of Basel insisted on such harmful and violent treatment that they finally gave up hope and ordered it amputated. Luckily Erasmus arrived just in time, countermanded the order, and sent for Paracelsus, whom he knew very well by reputation. The messenger found him sulking in a small town outside Strassburg, and the magic name of Erasmus brought him post-haste to Basel. The injured foot responded at once to a sane and moderate treatment, and Paracelsus followed up this .first triumph by curing Erasmus himself of a number of stubborn ailments which had bothered him for years.

After a few days in Basel Paracelsus looked about him and decided that if he were ever to be the medical Luther, this was the time and place to begin. The town was favorable to change of all sorts. The university had a constitution which forced it to yield to some extent to popular opinion. And he had on his side two of the most influential men in the city. He took Erasmus and Froben into his confidence, won them over to his great dream of a medical reformation, a union of medicine and alchemy. The three put their heads together to see what could be done.

The office of city physician, which carried with it a chair of medicine in the university, happened to be vacant, and Erasmus and Froben were able to persuade the city council that by appointing Paracelsus they would get a very good doctor for the place and at the same time would deal a telling blow on the conservative professors of the university. The Protestants on the council saw the point at once, and suddenly Paracelsus found himself in power.

He drew a deep breath and plunged into the battle. This was his great opportunity, and he wasn't going to let it slip by any prudent consideration for his colleagues' feelings. He resolved to dramatize his new doctrines before the people themselves. Perhaps if he won enough popular support, the doctors themselves would come around to his side. So presently a notice in large letters appeared on the door of the city hall: "The Famous Doctor Paracelsus, City Physician, will speak at High Noon to-morrow in the Town Square upon the New and Marvelous Light of Medicine. He will also touch upon the Ignorance, the Avarice, and the Strutting Vanity of the Doctors of Basel."

The people of Basel loved scraps, and this looked like a good one. The news spread rapidly, and long before noon the town square was crowded. The townspeople came in a mass; the peasants moved in from the countryside. Invalids and cripples dragged themselves from their beds to stare at the platform before the City Hall where a charcoal furnace glowed brightly with two strong men working its bellows. Exactly at noon Paracelsus appeared. He was dressed in a sweeping black silk robe trimmed with red. His hat was black and gold. He wore a long sword and carried an ebony staff. Behind him walked a page carrying two large books bound in leather.

For a moment he faced the crowd in silence, then strutted up and down the platform, sweeping the flagstones with his robe, showing off his staff, his sword, and his regal stride. Then he stopped, tore off his hat and threw it savagely into the audience, slammed his sword on the pavement, broke his staff over his knee, stripped off his robe, rolled it into a crumpled ball and sent it after the hat. He advanced toward the crowd bareheaded, in a plain gray jacket, sleeves rolled up to the elbows.

"Thus," he screamed in his shrill voice. "Thus should a doctor appear before his patient—to cure by knowledge, not by fine clothes, by science, not by gold rings and jewels."

He motioned to the page who handed him one of the books. With a furious gesture, Paracelsus tore it in two and threw it on the furnace. It blazed up in a burst of yellow flame and black smoke.

"*That* was Galen," he shouted.

The second book followed, and a second burst of flame rose up.

"*That* was Avicenna," shrieked the heretic doctor. "Old bloodless words. Vain mouthings of ignorance. Latin sounds meaning nothing. From these books your doctors get their Latin for diseases they know

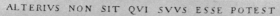

### PARACELSUS

Christened Aureolus Philippus Theophrastus Bombastus von Hohenheim, he applied chemistry to medicine and was the first modern doctor.

nothing about and their Greek for diseases they never heard of. Gray-bearded frauds, old wormy motheaten sophists, lousy pretenders with their fine clothes, their long steps, their Latin to hide their ignorance. They cling to the rich like leeches and let the poor die like flies. They make a disease out of nothing but a pain in the belly from too much eating. And when there is a real disease, they fly from it afraid for their reputations. Their cures are worse than the illness. They burn the flesh with hot irons, give black draughts which tear at the bowels. Their plasters raise blisters as thick as a hand. Then they go back to their snug studies, thumb over Hippocrates, that old Greek, and Galen, that old Roman, and count the golden coins they've stolen from your pockets."

The crowd listened in silence, but its silence was louder than cheers. The sick crowded the platform, looking up with dim worshiping eyes while Paracelsus told the story of his travels—the marvelous drugs he'd made by alchemy; the mercury which cured French pox; zinc ointment which cured sore skin; poppy juice from the East which veiled the most savage pain in a thick mist of dreams. And darker, less definite secrets—the Archeus which lives in the stomach and separates food from poison. Elemental spirits, gnomes of the inner earth, salamanders which live in the sphere of fire halfway to the sun.

Incubi and Succubi, half-souls hatched out by witches. Homunculi born of human seed in the warmth of the alchemical crucible. Dark matters these, far from the beaten track of medicine. But the beaten track of medicine was a small barren circle. Paracelsus had broken the restraining bonds and flown off at a wavering tangent into outer space where the secrets of life lay hidden still in darkness. He brought back much which was weird and foolish, much black magic which had haunted the dusty corners of the human mind since the world began. But he also discovered things of real value, and best of all, he'd broken the bonds of Galenic dogmatism.

When the flow of an oil well grows less and gradually dies away, a silent slow-moving man drives up in a rubber-padded truck with tanks of fretful nitroglycerine sloshing under the seat. He has to be a brave man, for he knows that if his load blows up, there'll be no attempt at a funeral. He carefully fills cylindrical cans with the uncertain stuff, lowers them into the well, drops a sharp-pointed steel weight, and steps back to feel the solid earth shake like jelly under his feet. Perhaps the well is ruined for good, the costly pipe shot up like water from a hose. But perhaps the oil flows again. At any rate, without this drastic step the well would be useless and barren.

Paracelsus was the well shooter of Sixteenth Century medical chemistry. He was far from slow-moving and silent. His words were his nitroglycerine. With shattering phrases, weird theories, untried cures, he smashed forever the barren system of Galenism. Medicine and chemistry sprouted like mushrooms after rain. Much of the growth was poisonous and unhealthy. But there was growth, and before there'd been nothing but words, dry dust, and decay.

The doctrines of Paracelsus started the bitterest controversy medicine had ever seen. Chemical medicine it was now, iatrochemistry, for Paracelsus had wedded the two sciences, and for a hundred years they went hand in hand. From its starting point in Basel, the conflict went all over Europe, running parallel with the religious fury of the Reformation. Universities were divided; old-fashioned doctors banded together for defense, fighting by boycott, libel, trials for black magic, even assassination. The spirit of the age was not calm and judicious. In medicine, as in religion, the old doctrines fought back savagely with any weapon which came to hand. Paracelsus soon felt the penalty for starting such an uproar.

For a year or two he had the upper hand in Basel. Behind him were Erasmus and Froben, the reforming clergy, and the rich townsmen he'd cured. But, tactless as always, he made more enemies than friends. The physicians, of course, were solidly and passionately against him. The apothecaries joined in the battle, enraged at the regulations which Paracelsus instituted to test the purity of their drugs and resenting the new doctrines which substituted cheap chemical remedies for the costly infusions of rare herbs brought from the East. These two groups would have been able to do little alone, but they set out to enlist allies. Paracelsus gave them plenty of talking points. Trade after trade he antagonized. He denounced the tactics of the lawyers. He'd go into a goldsmith's shop and tell at a glance which metals were adulterated. He detected commercial trickery of all sorts; he knew the tricks as well as the shopkeepers did. In the language of modern business, he "impaired confidence," and soon every guild in the city was bombarding the municipal council with petitions to get rid of this terrible doctor who was wrecking the fabric of secret agreement and mildly fraudulent compromise so essential to the commercial *status quo*. The council looked around for a pretext. It wasn't long before it found one.

A certain canon of the cathedral, wealthy and a solid rock of orthodoxy, fell sick of an unknown illness. He tried all the regular doctors without improvement, drank endless nasty syrups, took numerous medicated baths, and got sicker and sicker all the time. But he wouldn't call on Paracelsus, even though his friends urged him to. Better call the devil himself than one of his human playmates. But one day he saw death looking in at the window. He screamed for his attendants—even a canon feels a bit nervous about death—and told them to offer Paracelsus a hundred gulden for a cure. The heretic doctor arrived at once and set to work with his simple remedies. Perhaps he was helped by the patient's firm belief in his diabolical inspiration, but at any rate in three days the canon was well on the road to recovery. Paracelsus asked for his hundred gulden.

The canon looked up from his pillow, weak but quite returned to his senses. A hundred gulden for this henchman of Luther, this black magician! A hundred gulden of the money of Holy Church! A hundred gulden indeed! "Here's six," he said. "I know what my life's worth better than you do."

Paracelsus raged and swore; reached down deep into his ample stock of bad language; demanded his just and agreed payment. He'd cure the poor for nothing, but this fat parasite, this round-bellied Romish bloodsucker, would have to pay the full amount. Hadn't he saved his worse than useless life? Hadn't he given him a few additional years in which to trick the simple-minded of Basel? He'd go to law about it if it was the last thing he did in this world.

The canon smiled contentedly. He knew more about the law than Paracelsus did, and perhaps he knew the insecurity of the doctor's political standing. The case was taken to court and decided against Paracelsus. The municipal council saw a chance to unite the forces of orthodoxy and aggrieved commercialism against its troublesome doctor. It went so far as to rebuke Paracelsus for asking such a fee for a few visits and a little cheap medicine.

Paracelsus probably didn't care about the fee; he had plenty of money at this period. But the rebuke made him fighting mad—even madder than usual, which was saying a good deal. And he lost the last remnants of his tact, which was also saying a good deal. He wrote a pamphlet in his best style against the judges, denouncing them for deciding from the depths of their ignorance a matter about which they did and could know nothing whatever. The pamphlet was scattered about the streets, and the town was in an uproar.

It seems incredible that a matter as remote from public interest as the theory of medical practice should be the cause of a popular controversy, but at the time of the Reformation, religion and medicine were largely identified in the popular mind. Each had its orthodoxy and each its heresies. Paracelsus was a partisan of Luther. In attacking Galenism he was also attacking the doctrines of Holy Church. If he'd been a more tactful politician, he could at least have relied on the support of the Lutheran party, but he attacked the small tradesmen who formed the backbone of the Reformation as fiercely as he did the Church herself, applying as violent language to fraudulent apothecaries as to the ignorant Galenic physicians. He failed to learn the first lesson of the successful reformer—to appeal to the prejudices of the party you consider least harmful.

Before the pamphlet had been on the streets an hour, Paracelsus realized he'd gone too far. Erasmus and Froben couldn't protect him now. The judges and city council were furious. The doctors and apothecaries clamored louder than ever. The Church and the university were hot on the trail. A warrant was issued for his arrest and exile on an island in Lake Lucerne. He was warned just in time and fled over the border into comparatively friendly Germany.

The rest of his life Paracelsus spent on the road, curing a prince here, a peasant there; sometimes alone, sometimes followed by a group of enthusiastic disciples, but always observing, learning, raging against the doctors, the Church, and the established order in general. He died at Salzburg in 1541, a homeless wanderer, almost an outcast, but he died with the satisfaction of knowing he'd dealt the doctrines of Galen a blow from which they'd never recover.

The conflict he started continued after his death and grew into an intellectual battle involving everyone in Europe with the faintest interest in science or medicine. There were Paracelsans and anti-Paracelsans. He was denounced as a quack and glorified as a divine healer, the German Hippocrates. His followers carried his theories to dangerous extremes, dosing their patients with violent drugs such as sulphuric acid and bichloride of mercury and killing them off even faster than the Galenists did. The mystical side of his teaching inspired another group which saw in him a prophet of the dim gods and demons of the ages before human memory. The Oriental words with which he was fond of clothing his mystical theories became a secret language, and the name Paracelsus gained a place in magical incantations which it retains to the present day.

For several reasons it's hard to put a definite valuation on the chemical and medical discoveries of Paracelsus. Doctors and scientists of the Sixteenth Century didn't like to publish their discoveries, and when they did publish, they were apt to speak in a mystical language designed to carry meaning only to the initiated. This wasn't only because they wanted to keep the benefits to themselves. They had a real conviction that knowledge of Nature was a weapon too powerful for the unenlightened to possess. Paracelsus was almost the worst offender in this respect. It was commonly believed at the time of his death that his most important secrets had never been made public. The accounts of some of his famous cures seem to indicate that he knew more than he pretended to. And what he did put down in black and white is written in such difficult language that it's hard to follow the meaning.

Another confusing factor was the prevalence of literary forgery. Since an old book was considered more authoritative than a new one, the writers on alchemy were apt to date their works back a few hundred years and sign them with the name of some illustrious character of the past. The reputation of Paracelsus suffered from this in two ways. Various later forgeries were attributed to him, hurting his reputation when discovered. And a good part of his genuine discoveries were copied down after his death and attributed to a non-existent monk named Basil Valentine who was supposed to have lived much earlier. This made him look like the worst sort of a plagiarist, and only in recent times was he completely cleared of this charge.

His general theory of chemistry, on which he based his operations, sounds extremely fantastic to us. It contains a great deal of mystical trash which has no place in science. It confused the situation by adding three "principles," salt, sulphur, and mercury, to the four "elements" of Aristotle. In general, it was little improvement on the numerous theories which preceded it. But on certain specific points Paracelsus made definite advances. He believed in the possibility of transmuting metals, but didn't think it could be done with the existing facilities. This is just what modern science thinks about it. He made several valuable distinctions which stood for a long time—divided salts into "alums" and "vitriols," for instance, according to whether they had a metal or an "earth" for a base. Not until Sir Humphry Davy discovered that the earths were metallic oxides did this distinction fall.

He was the first to use the word "zinc" and the first to describe the preparation of the metal. In medicine he introduced a long list of drugs which may have been in use before by unorthodox physicians but which had never entered the practice of the profession in general. Most of them were chemical in origin. He used mercury compounds for syphilis, zinc ointment for skin diseases, and many other metallic preparations, a large number of which are still in use to-day.

But it isn't on his definite scientific discoveries that his fame depends. Paracelsus was a reformer, almost a prophet, and seldom do either reformers or prophets contribute anything definite to the knowledge of their day. They are leaders, rallying points. They see broad principles and fight to open the way for more productive but less courageous minds. Luther was the most conspicuous public character of the Sixteenth Century. His influence on the period was greater than that of any other man. But he was a poor theologian and a worthless philosopher. His importance came from the vigor with which he acted on his strong conviction that the Church needed housecleaning.

Paracelsus was the Luther of science. He was the reformer who blasted science out of its habit of looking back to the superior knowledge of the past. He believed, as we do now, that knowledge comes from accurate and unprejudiced observation, not from doctrine. Like Luther, he had to contend with the universal medieval conviction that the human race has little to look for in the future and had better study with reverence the all-wise revelations of the past. It wasn't only the Church which believed this.

Science and medicine thought the same way; believed that if you searched through the dim pages of antiquity you'd learn more than if you observed the pages of Nature which lie open before your eyes.

But after Paracelsus all this changed. The closed circles of alchemy and Galenism, like the closed circle of Catholic doctrine, burst in fragments and out poured an avalanche of new ideas, some of them weird and eccentric but others sufficiently sane to form the firm basis of modern medicine and chemistry. No longer did the alchemists give their whole attention to the pathetic search for the Stone. They had a better object now, to prepare medicines and pry into the chemistry of the human body. No longer did the doctors weigh one false assumption of antiquity against another equally false. They could draw on the vast accumulation of useful knowledge which the furtive alchemists had been building up for centuries.

So when we put iodine on an infected finger or watch some friend being brought back into the world by insulin or diphtheria anti-toxin, let's thank Paracelsus for giving the first rude push which started medicine on the task it has done so well. He hadn't the logic of Roger Bacon, or Bacon's level sanity, but he started the invaluable revolution which Bacon had failed to start two and a half centuries before.

# The First Bishop of Science

# CHAPTER V
## The First Bishop of Science

THE world does its thinking in spurts, with much noise and excitement—then rests for a century or two happily confident that it has found at last a firm philosophy of life. These "thinking periods" are painful to live in. The old order fades away into the past, and it looks very attractive as it dissolves in memory. The new order comes on with alarming quickness. It is hard, shining, harsh like a new and unfamiliar machine. It has gathered no sentiments, no warm traditions, and men look at it with fear, wondering how their spirits will avoid its sharp edges, how they can be happy without the comfortable old ways of life which this nascent monster will destroy so cruelly.

Such a period was the Seventeenth Century. Nothing stood firm. The Papacy was shaken to its foundations; Protestant sects sprang up like mushrooms, and religious warfare was the order of the day. The Divine Right of Kings was shaken too. Not so profoundly, but shaken nevertheless. And—most important of all—in the Seventeenth Century were celebrated the first triumphs of the Scientific Method, that calm, skeptical habit of thought which has grown gradually until it now dominates completely, for better or worse, the mind of civilization.

★    ★    ★

With a favorable breeze pushing it gayly into the harbor, a trim little ship approached the wharfs of Plymouth. The sailors ran about madly; sharp words of command crackled from the quarterdeck; sails flapped; men shouted. It looked like confusion undirected, but the serious, sad-looking young man who stood amidships knew better. He knew each movement was planned, each shout meant something essential; that officers and men alike had one object alone—to dock the ship safely. "How fine," he reflected, "to know what you're doing and have no thought beyond your task. It takes skill and judgment to manage a ship but how directly, how simply the problems are solved. How I envy these men! How I wish I were one of them! They see their problems so clearly. But they are sailors and I am Robert Boyle, son of the great Earl of Cork. And all England is lost in the tumult of civil war."

It was the year 1644, and the Puritans and Royalists were at one another's throats. The Cavaliers were riding the moors, and with them two of Robert Boyle's older brothers. The Roundheads held the towns of the south, the heart of England, and the battle of Marston Moor had just given them the advantage in the north. But nothing was settled. Armies marched this way and that across the country. Families were divided. Town fought against town, and village against village. The Irish were up in revolt. The Scotch flooded down from the north, not quite sure what they wanted but anxious to get into the fight. It was a tough country to return to after six peaceful years on the continent, and as he came ashore at Plymouth, Robert Boyle felt very much adrift and alone. He had no plans, no hope, no moorings. He'd lived for six years, from his thirteenth to nineteenth birthday, in the quiet places of Europe, studying under the care of a tutor, lavishly supplied with money from the rich resources of the earl, his father. But now that father was dead. England rang with war. His whole world had fallen, and the darkness was filled with cries and the sound of cannon.

And, worst of all, Robert Boyle didn't know which side he was on. He was pulled back and forth by conflicting loyalties. Brought up, as customary with boys of his high birth, to worship and trust in the Church, the King, Morality, Faith, all the interlocking traditions which buttress the *status quo,* he was drawn by the forces of early training and heritage toward the Royalists who were defending these traditions against change and destruction. But on the other hand, his brain told him that tradition was seldom the best guide. His six-year residence at Geneva and other Protestant centers had made him sympathetic with the Puritan leaders who wanted a new deal in religion and politics. And he knew that a new deal would have to come.

Thus, at the age of nineteen, Robert Boyle came face to face with the problem which troubled him all his life. Whether to follow his heart and the Cavaliers or his head and the Puritans. Deep within his being he longed for the calm, the tranquillity, the unquestioning faith of the past. But his brain was too alert, too inquisitive, too original. It told him that the past was gone forever; that religion had broken from its safe moorings; that from now on men would trust less in faith, and more in reason. The world was changing rapidly. Boyle's heart was sorry to see it change. But his head after all was stronger, and it made him a somewhat unwilling leader of the modern, the scientific age.

Robert Boyle was born in 1626. He was the seventh son of the Earl of Cork, the leading peer of Ireland, and one of the richest Englishmen alive. He was a

sickly child, and he stammered terribly. When he
was sent to Eton at the age of nine, his weakness, his
stammer, his private tutor and private meals intensi-
fied his natural shyness and made him into a timid
boy much more fond of his studies than of the famous
"playing fields" which two hundred years later were
to be thanked for winning the Battle of Waterloo.
During his three years at Eton his health grew rather
worse than better, and so in 1638 he was sent to travel
on the continent with his seventeen-year-old brother
Francis and a French tutor M. Marcombes.

It was a common thing to send young noblemen to
the courts of the continent for a general polish and
broadening, but this time the motives of the old earl
were very different. He'd seen too many of his
numerous sons grow up to "drinking, dicing, and
wenching" at the merry court of England. He vowed
his two youngest should be educated in a very different atmosphere. So Geneva it was instead of
Paris, Vienna, or Madrid. Dour Geneva, lair of
shadows, stronghold of grumbling Calvinism, "where
they would be in no danger from conversations with
Jesuits, friars, priests, or any other persons ill-af-
fected toward their religion, king or state." Poor old
earl! He saw the changes coming over the horizon
which would upset the loyalties he had served so long

and so faithfully. He thought that at Geneva at least there was peace and security. He made the all-too-common mistake of confusing social dullness with stability. Geneva was certainly dull to courtiers' eyes, but it was also staunchly Puritan, and Seventeenth Century Puritanism was a quarreling, questioning faith which attacked every established doctrine from the sanctity of marriage to the Divine Right of Kings. A poor place to send a young boy if you wanted him to keep his mind placidly in tune with the past.

It was a gay England the two boys left behind them, an England enjoying perhaps consciously the last few pleasant years before the oncoming storm broke over its head. The court was brilliant, and the old earl and his numerous sons and daughters were in the very midst of it all. He was very close to the king himself, and when Roger Boyle, Baron of Broghill, married charming Margaret Howard, elegant King Charles himself forgot his political troubles for the time and presided with gayety and good humor. There were plays and music and dancing. And Sir John Suckling wrote a poem in honor of the bride which contains three lines which will live forever:

> "Her feet beneath her petticoat
> Like little mice stole in and out
> As if they feared the light."

It was a gay England. And it would be pleasant to think that Robert Boyle regretted leaving it. But he probably did not. He was being sent to Geneva for moral buttressing, but never did boy need it less. He had enough morality already to keep him bothered the rest of his life. His letters home sound like those of a somewhat unworldly bishop. What would such a boy be doing at the merry court of the Stuarts?

Geneva was a curious city in those days although perhaps no more curious than it is to-day with the League of Nations cluttering its streets with the cranky idealists of all the world. It made a great show of toleration, which meant, since the Protestants of various shades were the only people in need of toleration, that every sect, from dictatorial Calvinism, already something of an established church, to wild and furious Anabaptism, was represented in this city of dreadful preaching. Religion was almost the only topic of conversation; it was talked on street corners, in shops, and in parlors. Every other man was bent on converting his neighbor to his own particular view. Sunday slopped over into the week, and at every hour of every day some earnest minister could be heard haranguing his flock. Geneva was a religious clearing house. Religion was the mainspring of its life.

Of course this affected Robert Boyle. It couldn't

help doing so. He was not strong physically, and weaklings are singularly prone to religious introspection. His tutor, M. Marcombes, seems to have been some sort of Huguenot. He favored perhaps, certainly didn't oppose, Robert's Puritan leanings. He wasn't alarmed when a sudden thunderstorm threw him into a fever of apprehension lest he be caught unprepared for the Judgment Day. Well-brought-up children in those days were expected to dwell much on their chances of bliss or hell fire.

When, after eighteen months in Geneva, the party moved on to Italy, Robert's mind was already set in a Puritan mold and proof against the theological blandishments of Rome and the well-advertised fleshly temptations of Italy. His account of that fascinating country is written in such a serious vein that the reader must constantly remind himself that this was a boy of fifteen, not a graybeard of sixty who looked on the amusements of the Italians with such disapproving eyes and found on every street corner still another proof that the Pope was no fit leader for genuine Christianity.

The old Earl of Cork may have worried a bit over his sons in wicked Italy, but needn't have. Not about Robert at least, for the young Puritan had already developed the typically Puritan habit of getting more intense pleasure out of resisting temptation than of yielding to it. How the earl's elderly virtue must have rejoiced at this letter from his son in Florence:

> "When Carnival was come (the season when madness is so general in Italy that Lunacy does for that time lose its name) he [Robert] had the pleasure to see the tilts maintained by the Grand Duke's brothers and be present at the gentlemen's balls. Nor did he sometimes scruple, in his Governor's company, to visit the famousest Bordellos, whither resorting out of bare curiosity he retained there an unblemished chastity and still returned thence as honest as he went thither, professing that he could never have found any such sermons against them as they were against themselves: the impudent nakedness of vice clothing it with a deformity description cannot reach and the worst of epithets cannot flatter."

—No, there was no danger of Robert's falling into the ways of certain of his older brothers.

The Boyles and their tutor made a short visit to Rome and then returned to Geneva. It was now the spring of 1642 and alarming news began to arrive from England. King and Parliament were laying aside all pretense and preparing for civil war. The Irish had risen, as they always did when the English were busy elsewhere, and the old earl had hastened off to protect his estates from the "mere Irish" on one side and the Catholic Lords of the Pale, "no better than Irish," on the other. Naturally the supply of money failed. Some was sent, but it never arrived. There was nothing to do but stay in Geneva, where M. Marcombes had his home, live as cheaply as possible, and pray that England would calm down, or that the right side would win.

The right side? Robert Boyle began to wonder which *was* the right side. He was only seventeen but already he began to suffer from that conflict of heart and head which bothered him all his life. By birth and training he was a Royalist son of an earl whose long, active, and successful life had been spent serving the crown, whether in the person of Elizabeth, James, or Charles. He had been brought up an Anglican, been trained to feel that the perfect philosophy of life was the ancient theology of the Christian Church, purged happily of such evil growths as the Papacy and the Mass. And in matters of government he had been brought up to worship that marvelous structure of tradition, non-logic, and compromise which had hitherto kept the English state on a more-or-less even keel.

But Geneva—this was no place to go if you wished your mind to crystallize in the ancient forms. Geneva was a fermenting vat of repellent but vigorous thought. Here was Puritanism in its thousand forms, feeling its strength and preparing to sound its discordant bugle call and announce the birth of a new age.

In Twentieth Century America, Puritanism is the dead hand of a very unpleasant past. There is no joy in it, no light, no progress or hope of progress. It lives in moldy laws and unreasoning, unfeeling conventions. But in the Seventeenth Century this was not so. The Puritans were the innovators. They stood for liberty of thought, for freedom of personal belief, for the right of individual judgment. Against them were arrayed all the forces which the modern Puritans cherish as allies—the established order, the king, the courts, rural ignorance, and social conservatism. To be a Puritan was to combat all these reactionary forces. To champion the individual against Church and State. To pit the liberalism of the city against the incurable mental stagnation of the countryside. To fight for self-government against the Divine Right of Kings. Puritanism shortly became more illiberal than ever its opponents had been, but that's the way with all successful heresies. Its spirit at the beginning was that of violent revolt against the past.

This questioning spirit Robert Boyle absorbed at Geneva, and it counteracted the traditional loyalties which otherwise would have made him a natural Royalist and supporter of the king. But neither tendency won a final victory. All through his life his active skeptical mind forced him to find new truths, map out unknown intellectual country, while his conservative spirit mourned inconsolably over the precious traditions which these discoveries destroyed.

In 1644 Robert Boyle sold some jewels and managed to reach England. The Civil War was at its height, and he was confronted by a dilemma. His deepest unreasoning feelings drew him toward the king. And his mind, churning with new ideas from Geneva, drew him toward the Parliament. It is hard for a modern American to imagine a man drawn to the Puritans by reason and drawn away by loyalty to tradition. Nowadays the positions are always reversed. But such was the situation with Robert Boyle. The Puritan party was new, innovating, experimental; it was progressive in abstract thought as well as in politics and religion. Such a party appealed to the mind of this serious, pale youth with his inquiring mind and his background of Genevan theology. But his deeper feelings, the feelings that do not need words or logic to express their decrees, told him that to join the Parliament would be treason to his class, his ancestry. Was he not the son of the Earl of Cork, the Elizabethan nobleman who'd labored to build up the power of the Crown in savage Ireland? Were not his brothers fighting with the Cavaliers? Had not King Charles himself, elegant and gracious, officiated at his family's numerous weddings? If he joined the Parliamentarians without in some way quieting this unreasonable conscience of his, he looked forward to numerous guilty regrets.

Luckily he didn't have to decide for himself. After all, he was only nineteen and too weak physically to be of much use to either army. So as soon as he entered London his favorite sister, Lady Ranelagh, a staunch but moderate Puritan, gathered him into her Parliamentary circle and insulated him carefully against any overtures from the other side. Robert was surprised to learn that the Cavaliers were not noble and patriotic gentlemen but a pack of thievish looters with no morals or principles. The king was no longer the just and kind monarch he used to be but an unreasoning tyrant who employed most evil advisors. The elder Boyle brothers were fighting for the king with marked lack of enthusiasm, and the most important of them, Broghill, was actually on the Parliamentary side, although he had made Cromwell promise that he would send him only to Ireland to replace the earl his father who had just died.

Robert took a general view of this information in search of something to exhibit to his intractable conscience. He found what he was looking for. "Ah," said this good young man (it must be admitted that Robert was loathsomely and revoltingly good)— "Ah, here it is." And he wrote down in his diary "that there were in the Royalist Army beside the excellent king himself diverse eminent divines and many worthy persons of several ranks; yet the generality of those he would have been obliged to converse with were very debauched and apt, as well as inclined, to make others so." And he thanked his lucky stars that he was living in the religious and virtuous household of his sister instead of riding the countryside with the dangerously unvirtuous Cavaliers.

Whatever might have been the moral advantages gained by living with Lady Ranelagh, there was a practical advantage also. The fortunes of the king were waning. Most of the country was in the hands of the Parliament, including Dorsetshire which contained Robert Boyle's manor of Stalbridge where he proposed to live until the war was over. Through her acquaintance with high Parliamentary leaders Lady Ranelagh managed to get her brother a safe

conduct through the Parliamentary lines and a
bodyguard of soldiers to take him to his own door.

Stalbridge is a small place in a remote part of
Dorsetshire far from any good-sized town. Stalbridge
Manor was an ideal hideout for anyone who didn't
intend to take an active part in the somewhat hectic
political and religious doings of the times. Many
sensible Englishmen were thus in retirement, wondering at the spirit of extremism which had taken
possession of their country and carefully refraining
from committing themselves to either side. Robert
Boyle was one of these. He considered warfare a poor
way to settle a controversy, and he knew very well
that, favorably situated as he was with powerful
friends in both camps, an attitude of neutrality
would be the safest in the end.

When Boyle took possession of Stalbridge Manor
he was only nineteen years old, but never did a more
serious young person look with supercilious disapproval at the habits of humanity or seek to draw
morals from everyday events. This was the period in
which he composed his "Occasional Reflections,"
little essays of such intense moral smugness that few
modern readers can glance at them without anguish.
Such syrupy goodness, such self-confident superiority
to human weaknesses! The titles of these little sermons, which outline the text after the fashion of the

day, show enough of the spirit to spare us the necessity of reading the rest.

We can see the young philosopher, grave and handsome in a pale way, walking about his pleasant estate and observing with a critical moral eye the life of the quiet countryside. It is early morning and some feathery pinkish clouds hang in the sky. Boyle looks at them with pleasure—as who would not?—but they set the moral mechanism in his mind to grinding out a little sermon. He takes pen and paper and writes a "reflection," calling it:

"Upon the sight of some variously colored clouds.—Gaudy and glittering favorites are vain and short-lived shadows, often destroyed by the very hand that raised and decked them out for the public eye to gaze at and admire."

—No wonder when he looked up again the clouds had faded.

A little farther on he hears a lark and sees the bird disappearing into the sky, leaving behind him a thin trail of music. Boyle watches, sensing that a moral is about to suggest itself. And so it does. The lark comes back to earth, lights on a clod, and blends so perfectly with its surroundings that until it moves to pick up a worm, it is almost invisible. Boyle takes his pen and writes another "reflection," pointing out that the lark, although it sings as if it were really part of the brilliant sky above, can nevertheless descend to earth in search of a vulgar worm and become once more a part of the earth's mean surface.

> "Upon the mounting, singing, and lighting of larks.—Hypocrisy is odious; but that circumstance does not excuse the open libertine; it is the pretense of religion and virtue we despise, and the best way of avoiding such pretense is really to practice them."

—That lark probably moved off Boyle's property the very next day.

But the occupation which was most productive of "reflections" was angling. Boyle, like most English country gentlemen, was devoted to this placid sport, not only because it made few demands on his small stock of strength, but because it gave him a chance to meditate. He writes much of "Angling improved to spiritual uses." Here are a few of his titles:

> "Upon the being called upon to rise early on a very fair morning.—We should not be discouraged from endeavoring to reform others by their unwillingness to listen to us, or even by the dislike they may at first conceive against us."

> "Upon fishing with a counterfeit fly.—Men are often tempted to do wrong even by a mere show of advantage."

"Upon a fish's struggling after having swallowed the hook.—Ill-gotten advantages always bear their sting along with them."

"Upon a fall occasioned by coming too near the river's brink.—It is dangerous to go too near the very verge of what is lawful."

It was probably fortunate for Boyle that angling is a solitary sport. Even in that day of preaching and moralizing, he'd probably have found few real anglers who liked to go fishing with a man in such a state of mind.

These "Occasional Reflections" were not written for publication, but they were published nevertheless and gained a certain amount of popularity among the moral-minded. So much so that the terrible Dean Swift saw fit to satirize Boyle's pompous style and sickish moralizing in his well-known "Essay on a Broomstick, after the manner of the late Robert Boyle." Needless to say, the "Essay on a Broomstick," although becoming a bit dim itself, will live long after any of its models.

It is hard while reading these "reflections" to believe that they were from the pen of a man who was to become a great scientist. They are written skillfully in what was then a much admired style, but beyond this they have no merit. Most clergymen can do as well and many much better. They contain no idea, no illustration which wasn't old a thousand years before Boyle went to live at Stalbridge. But even if they have no merit of their own, they show some very important things about Boyle's character.

Robert Boyle was never in very good health, and his incurable stammer together with his constant illnesses kept him from most companionship. Deprived of their natural outlet in sport and the innocent or not so innocent diversions of normal young men, his powerful mind and ambitious spirit turned to other fields. The less normal a man is the more he looks for ways to distinguish himself, for he feels that unless he definitely excels in some respect, he will fall below the general level of his normal companions.

To a young man suffering from physical handicaps there are two roads open to the pleasant country of self-esteem—superior righteousness and superior mental accomplishments. Both roads are well traveled, and the types on each are familiar. Toward righteousness travel the gentle saints who inspire by courageous suffering, the homely, angular old maids, cheated by life, the half-lunatic Carrie Nations, and the fanatical reformers whose single desire is to make others lead the dull unhuman existence forced on them by some mental or physical disability. Toward intellectual eminence travel the physically han-

dicapped whose whole energies have been diverted
into their brains. The eunuch Paracelsus, the blind
Milton, the epileptic Dostoevsky, and the hunchback
Steinmetz. It could probably be proved that much
of the good as well as most of the harm has been
done to the world by men whose mental or physical
disabilities have made them feel that if they do not
impress their personalities strongly on society they
will be classed for good with the social waifs and
strays.

Boyle is remarkable among this varied company in
that he took the first road, decided it led to no good
destination, and turned back to take the second. To
judge by the "Occasional Reflections," he was, in his
early twenties, well on the way to becoming an in-
tolerable preacher and moral faultfinder with an in-
sufferable sense of his own virtue and a fanatical
hatred for the pleasant mild vices of life which his
health made it impossible for him to enjoy. But he
never went beyond this point. He never developed
from the preacher into the persecutor; never made
any great effort to force others into his way of life.
He remained virtuous enough; it isn't difficult for a
semi-invalid. But his righteousness became less and
less annoying until at the end of his life he was a
tolerant, slightly amused spectator at the pleasantly
dissolute court of the Restoration. What brought

about the change was this: He found an interest worthy of his powerful mind. Shortly after writing the "Reflections" he became absorbed in "natural philosophy," meaning physics and chemistry. And moral fanaticism can't live, much less grow, in the logical atmosphere of a laboratory.

Let's examine now his two great discoveries. We'll find a very different quality of mind from that shown in the distressing "Occasional Reflections."

By 1654 the Civil War had quieted down, Cromwell was sitting crownless on the throne, and Boyle had moved to Oxford in search of "philosophical companionship." He had been working since 1647 on various scientific problems but hadn't accomplished much. The times were too confused, supplies were hard to get, and money was scarce. But now the situation was better, and he began to make progress. The first thing he tackled successfully was the problem of the vacuum.

The fact that water, for no apparent reason, follows upward the piston of a pump had long bothered the "natural philosophers," and since they had no better way of explaining it, they developed the theory that "Nature abhors a vacuum," and consequently draws up the water to fill the space below the piston.

It was solemnly stated with great metaphysical subtlety that Nature, that vague abstraction so often

confused with God, was determined that no space
should exist unfilled with some kind of matter. Of
course, to the scientific mind, this is no explanation
at all. Nature isn't supposed to have emotions. But
in the Seventeenth Century, the scientific mind was
only half developed. The reasoning of the time was
strongly tinged with theology, and it seemed perfectly all right to allow Nature to have at least a
few of the irrational prejudices which the various
deities have enjoyed since the world began.

Of course most of the thinkers of the time realized
that the "abhorrence" theory was inadequate, but because they were half theologians, they tried to improve it instead of throwing it overboard entirely.
Theological minds always try to defend against all
odds the errors of the past. Descartes, the great
French philosopher and mathematician, wandered
off into metaphysics, maintaining that if a space existed with nothing whatever in it, the walls of that
space would be separated by nothing, and therefore
could be said to touch. And where is the vacuum?
Hobbes, slightly more practical-minded, declared
that the universe was entirely full, and illustrated
his theory as follows:

> "If a gardener's watering-pot be filled with
> water, the hole at the top being stopped, the
> water will not flow out of any of the holes in

the bottom; but if the finger be removed to let in the air above, it will run out of them all, and as soon as the finger be applied to it again, the water will suddenly and totally be stayed again from running out. The cause whereof seems to be no other but this, that the water cannot, by its natural endeavor to descend, drive down the air below it because there is no place for the air to go unless by thrusting away the next contiguous air it proceed by continual endeavor to the hole at the top, where it may enter and succeed in the place of the water that floweth out, or else by resisting the endeavor of the water downwards penetrate the same and pass up through it."

Boyle took a look at the various theories and decided something was wrong. He very sensibly dismissed the Descartes theory as a matter of words, not of facts, and set to work on Hobbes, whose theory had gained so many followers that they had a name, the plenists.

There were many things wrong with the plenist theory. It explained such things as the action of watering pots and such, but very little else. If, for instance, the watering pot were made in the form of a pipe thirty-five feet long, the water would not wait until a hole was opened at the top, but would flow out at once until its level stood between thirty-two and thirty-three feet above the bottom. Where had the air gone which this water replaced? Furthermore, if the same pipe were taken to the top of a high mountain, more of the water would flow out than at sea level. Why did Nature abhor a vacuum less at the top of a mountain than at the bottom? And if mercury were substituted for the water it never rose higher than some thirty inches. It looked as if mercury had more influence with Nature than water had.

Now before you can study a vacuum you have to capture one and domesticate it. Up to the time Boyle took charge of the problem this had not been done. The Torricellian vacuum at the top of a column of mercury is a very good vacuum indeed, but it's hard to manage and practically useless for experiment because there's no way of getting at it. In 1654 Consul Guericke of Magdeburg demonstrated the first air pump, a simple machine much like an ordinary bicycle pump but working in reverse so as to take air from a vessel instead of forcing more in. It had one disadvantage. It was so faulty in operation that it had to be worked under water, which naturally destroyed much of its usefulness. But it was right in principle, and Boyle gives Guericke all credit for its invention.

In 1658 Boyle went to London to look for an

"artist" capable of carrying out some improvements which he had devised for Guericke's air pump. He found a certain Robert Hooke and brought him back to Oxford. After months of work and trial the "pneumatical engine" was finished, and the experiments began.

It was a curious-looking machine which Boyle and Hooke constructed and it must have been a terrible thing to handle, but it worked, which was more than any previous device had done. The essential part was a brass cylinder with a piston operated by a gear wheel and crank. The large glass "receiver" was set in cement over the intake pipe and at the top was a brass cover which could be removed to insert the various objects on which Boyle proposed to experiment. There were no valves. A brass cock had to be turned after each stroke of the piston to keep the air from seeping back through the leather washer, and "sallet oil" was poured liberally over all working parts to lubricate them and discourage leaks.

As soon as the machine was finished Boyle set enthusiastically to work investigating the properties of the vacuum. The first thing he did was to prove the "spring" of the air by placing partly inflated bladders in the receiver and removing the air from around them. The bladders swelled and eventually burst. Next he filled with water a U-tube closed at one end leaving a small bubble at the top. As the air was pumped out, the bubble grew until it filled the whole leg of the tube. The spring of the air was demonstrated. There seemed to be no limit to its spring.

At this point it looks as if Boyle became so fascinated with his new toy that he forgot science for a time and amused himself with innumerable interesting but useless experiments. He put everything he could think of into his vacuum to see what effect it would have on them. Fruit, human blood, lighted candles, boiling pitch, camphor, and gunpowder. And especially living animals. Many were the butterflies, sparrows, and mice who died for the sake of science and Boyle's curiosity. But he must have been a soft-hearted man after all, for once when he left a mouse overnight in the receiver, he made it a soft bed of paper scraps, and gave it a bit of cheese to console it during the long hours until morning. And when he found it still alive, he let it go free, saying it had suffered enough for any one mouse.

But in spite of these diversions, Boyle was a real scientist, which meant that he had a quantitative mind. It wasn't enough that the vacuum had a remarkable effect on many things. He must measure his vacuum accurately and find out whether it was the same as the mysterious space above the mercury of a barometer. This brought him to the crucial experiment, which was to place a barometer in the receiver and see if its mercury fell as the air was removed. If it did fall, it would prove that the mercury was held up by the pressure of the atmosphere, and the doctrine that Nature abhors a vacuum would be thrown on the scientific scrap heap with the constant speed of falling bodies and the central position of the earth in the universe.

It was almost with a feeling of excitement that the calm and collected Robert Boyle turned the brass cock and let the air from the receiver rush into the cylinder of the pump. The mercury in the barometer fell several inches. Nature had lost her last fully certified emotion. The "abhorrence" theory had fallen, and with it fell much of man's theological habit of mind.

From here it was an easy step to the complete formulation of Boyle's Law, the fundamental law of gases, which has not undergone any essential modification to the present day. It states simply that the volume of a gas is inversely proportionate to the pressure. It is so simple and so well known that we are apt to take it for granted, but never did a discovery lead to greater consequences. It was the first real fruit of the scientific method, the first conspicuous proof that the most intimate rules and regulations of the universe could be detected by experiment.

The discoveries of modern science require so much technical preparation and such elaborate apparatus that the layman seldom hopes to understand how they are made, but in the days of Boyle, the apparatus and calculation were so simple that anyone could understand them. The average schoolboy could have made all the momentous experiments on gases in the average woodshed.

Boyle took a long glass tube and bent it into a U-shape with one leg much shorter than the other. He sealed the short end and poured in enough mercury to fill the curved bottom, taking care to tip and shake the tube until the mercury in both legs stood at the same level. This proved that the air in the sealed leg was at the same pressure as the atmosphere. Then he pasted strips of paper on both legs and marked carefully the horizontal line passing through both surfaces of the mercury. He divided the strips of paper above this line into equal spaces so that any changes could be easily measured.

The object of the experiment was this—to see if a column of mercury long enough to double the pressure of the atmosphere on the air in the short leg would reduce the volume of that air by one half. The barometer that day stood at 29⅛ inches—rather remarkably low, but in the Seventeenth Century people were not particular about barometers. This meant that the air in the closed part of the tube was already under pressure of 29⅛ inches. If it took an additional 29⅛ inches to compress it to half its volume, the hypothesis would be proved, and the scientific method would have gained its first great victory.

The calm and collected Boyle actually *was* excited this time. He admits it grudgingly in his notes. He took a funnel and began to pour mercury into the open end of the tube. His assistant watched the closed end, and when the air in it was reduced to half its former volume, he motioned to his master to stop pouring. Boyle put down his funnel, and noted "not without delight and satisfaction" that the mercury in the open tube stood higher by the same magic 29⅛ inches. Boyle's Law was born, and Robert Boyle at the age of thirty-two had built himself a lasting monument. Most men do no more than cut their names on a slab of stone. Boyle carved his name and law on the resentful minds of fifty million schoolboys.

Boyle's work on the properties of gases and the furious controversies which followed—the theological-minded plenists were hard to silence—gained him a great reputation, but probably his discoveries in chemistry had more far-reaching consequences. The science of chemistry had taken great strides since Paracelsus put the medieval alchemists on the defensive, but its general theories about the constitution of matter were, if possible, more confused than ever. Paracelsus had turned the alchemists to the productive business of compounding drugs, but he hadn't told them much about the make-up of the materials they worked with. In fact the mystical passages and numerous contradictions in the works attributed to him encouraged a sort of inspirational chemistry which produced more noise than results. Since his death the chemists had discovered a great many new compounds and reactions and had made great advances in technology, but they knew little more about the rules that governed their experiments than they did before. The three "principles," sulphur, salt and mercury, which Paracelsus added to the Aristotelian "elements," earth, air, fire, and water, only increased the confusion, for the distinction of the elements as the constituents and the principles as the qualities of all matter was too hazy and difficult for any two men to agree on. As Boyle expressed it, "Methinks the vulgar chymists are like the ships which Solomon sent to Tarshish. They bring back not only gold and silver, but also peacocks and apes"—a remark which contained not only truth but humor, the latter quality very painfully rare in all his works.

When the modern chemist speaks of an element he means a substance which cannot be broken down by chemical means. The Peripatetic chemists who flourished before the time of Boyle meant something very different. The four elements, earth, air, fire, and water were the four fundamental constituents which, compounded together in varying proportions, formed all substances. There was a disagreement about whether an element could exist in a free and pure state. Some of the Peripatetics felt safer in saying that it could not. This headed off any skeptical person who wanted to see "pure fire" for instance, or "pure earth."

Unfortunately there were a great many substances hard to classify under any of these heads which defied all attempts to break them down into simpler forms. The metals were most intractable. Gold, for instance, could not be broken down, nor was it a pure element. The same was true of silver, mercury, and various other substances such as sulphur and certain precious stones.

Paracelsus saw these difficulties and decided that substances contained not only one or more of the four elements, but also varying quantities of what he called the three principles, sulphur, salt, and mercury, which he was very careful to state were not the definite substances we now call by these names. They were subtle, semi-material "attributes" which gave to mixtures of the four elements their various char-

acteristics. Sulphur was the principle of inflammability. It was contained in all substances which would
burn. Mercury was the principle of volatility and
fluidity. It was contained in all liquids and solids
which would give off vapors when heated. Salt contributed "fixity." It was the predominant quality in
all non-volatile non-inflammable substances such as
stone, quick-lime, and sand.

This was bad enough, but some of the chemists
went further. Van Helmont, for instance, a leading
chemist only a little before Boyle, insisted solemnly
that he possessed the "alcahest" or universal solvent
which could reduce all matter to the "primitive
atoms" common to all substances. It was an astonish-
ingly long time before anyone thought to ask him if
he really had this alcahest what kind of a vessel he
kept it in.

These various theories had been developing a long
time and had become exceedingly confused. No two
chemists agreed. Each one could point out flaws in
the theories of all the others. But the system of elements and principles was so vague and allowed so
many varying interpretations that to refute it completely was like trying to catch an eel in a muddy
pond on a dark night. The Peripatetics couldn't
agree on anything, but they had so confused the issue

that no one could disprove conclusively anything they said.

Here was a task worthy of the "careful and doubting Boyle" and a true test of the scientific method. The previous attempts to find a firm basis for chemical reasoning had failed because the chemists tried to improve on the theories of the past, and these theories were fundamentally wrong. There are no elements making up all matter, and any amount of modification of the four-element theory wouldn't improve things. A new start had to be made, and Boyle decided that he was the man to make it.

Here is where the scientific maxim, "State nothing which you can't prove," was first applied to chemistry. The early theories had inherited the four elements from antiquity. They tried to reconcile the facts with the theory. Boyle started with the observed facts and tried to find a definition for an element which did not conflict with them.

He reasoned something like this. "Most substances are compounds. We can prove this by separating them into other substances. Some, however, we cannot break down. So let's for the present call them elements, and if they continue to resist our efforts to break them down, we shall continue to call them elements. But once a substance is decomposed, it at once loses rank as an element and becomes a compound."

This, of course, is good reasoning and the kind of reasoning every scientific worker now applies to every problem. Start with what you have and try to build up something better, being careful not to be led astray by attractive theories which do not agree with all the facts. But in the time of Boyle it was something of a novelty. The fashion of the day was to pray for inspiration, take deep thought, evolve a perfect, complete theory, and then try to find facts to uphold it.

With immense labor and infinite pains Boyle set to work on the program he'd laid out for himself. He was a very rich man now, for the gradual quieting down of civil troubles had restored his estates intact, and he spared no expense for materials, apparatus, and assistants. His account of his work is tedious and badly arranged like all his writing, but his method shines through his words with a quite sufficient light of its own. It was very simple. He took each substance in turn, from human blood to iceland spar, and applied all known methods in hopes of decomposing it. Some substances yielded at once to heat and separated into "phlegm, oil, and ash," thus proving that they were compounds or mixtures. Some, like the various ores, required more compli-

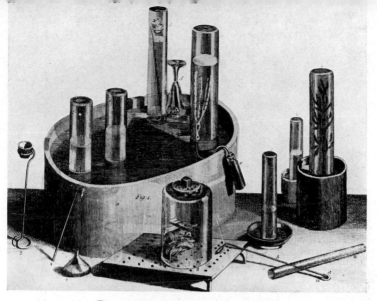

PRIESTLEY's pneumatic trough.

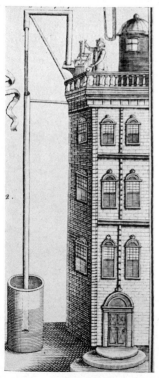

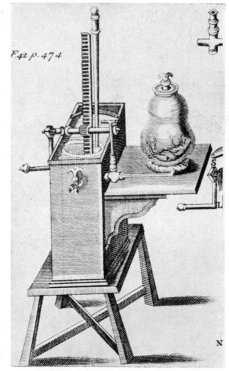

Fig. 1.

Fig. 2.

1. ROBERT BOYLE testing nature's "abhorrence of a vacuum."
2. A HUMBLE MARTYR to science—a mouse in a pneumatic jar.

cated methods. And some, notably the metals, resisted all treatments and remained unchanged. These Boyle placed in a class by themselves and for the time called them elements.

The results of these researches he published in a small book which appeared in 1661. It was called, with the pomposity of the age in the matter of titles, *The Skeptical Chymist, or Considerations upon the experiments usually produced in favor of the four elements and the three chymical principles of mixed bodies.* It isn't exactly what you'd call light reading, nor does it resemble the sharp and concise, if completely colorless, articles of the modern scientist. For tediousness, bad arrangement, and verbosity, the books of Boyle have few rivals. But this was the literary fault of the age; it was what people expected, and *The Skeptical Chymist* made a tremendous hit with the "philosophical" minds of the time. Boyle's conception of an element was what the world had been waiting for, a sort of guide post to lead the science of chemistry through the confusion which had been steadily growing since man first began to ask the why and wherefore of chemical phenomena.

Of course Boyle did not apply his method rigorously to all substances and discover all the elements. This was beyond the equipment of the period, and Boyle, who touched on every subject from the saltiness of the sea to the conversion of the New England Indians, had other things to think about. But the idea of a chemical element which can not be decomposed still remains the central concept of the science.

These two discoveries, Boyle's Law and the chemical element, concern very different problems, but they had one remarkable thing in common. Each was a successful attack upon a mistaken scientific doctrine which had been established so long that it was almost an article of faith. They were not made, like so many scientific discoveries, by a small-minded man who stuck to his work until he found out something important. Not at all. Boyle surveyed the intellectual scene, decided that the current theory rested on shaky foundations, and proceeded by reason and observation to build anew from the ground up. He was not afraid of the accumulated opinion of the past. He cleared away all dogmas inherited from the Middle Ages and built his structure on the solid bedrock of personal observation. This doesn't sound very remarkable to modern ears. All scientists work that way now, or pretend to. But to do it in the Seventeenth Century when the dead hand of the past was still heavy, and to do so successfully, raises Boyle far above the patient fact-gatherers into the class of Newton, Lavoisier, and Einstein.

This remarkable intellectual independence, this

freedom from the influence of the past, calls up a picture of Boyle as a freethinker, an innovator, anxious to call for a new deal and rebuild the world on a new and more rational basis. Nothing could be more untrue. Boyle's mind had in it two idea-tight compartments. On the scientific side it was rational, skeptical, unconvinced by anything but rigorous proof. On the other, the theological side, his mind was moral, conservative, unexperimental; he sighed inconsolably for the good old age of faith, which then, as now, seemed to have passed forever.

During a visit to London under the Commonwealth, Boyle wrote a letter which illustrates how he felt about the religious instability of the times. London was a parade ground of all extreme Protestant sects of Europe, and Boyle was disgusted with it. He unburdens himself thus:

"Few days pass here that may not justly be accused of the brewing or broaching of some new opinion. Nay some are so studiously changling in that particular that they esteem an opinion as a diurnal, after a day scarce worth the keeping. If any man have lost his religion, let him repair to London, and I'll warrant him he shall find it. For my part I shall always pray to God to *give us the unity of the spirit in the bond of peace.*"

The last phrase is significant—*give us the unity of the spirit in the bond of peace*. In modern language it means about this—"Let's decide what we're going to believe and shut up about it." This is a strange opinion for a scientist. Men of science, if they are good ones, never hope for final decisions. They know from long experience that the best theories fall, that new discoveries undermine the most solid doctrines of the past, that even the theories which remain standing will be pushed into the background by greater discoveries to come. They know that this applies to morals and government as well as to science. The real scientist does not hope for a final settlement of anything or expect one. He knows that the only permanent thing in the universe is change.

But Robert Boyle was not entirely a scientist; he was a transition phenomenon between a past which wanted to believe and a future which was to hope eagerly for a chance to doubt. When he was thinking on purely scientific matters he was clear, logical, he never took anything on faith; he tested every theory himself, and if he found it faulty, he said so without hesitation. But when he locked the door of his laboratory, when his furnace was cold and his "pneumatical engine" was quiet for the night, then he felt the tradition of a thousand years drawing him back to the peace, the tranquillity, the unquestioning faith which he feared the world had lost forever. He went to church and listened carefully to the sermons, hoping against hope for proof that the cherished doctrines of Christianity were on sounder foundations than his reason led him to fear. He mourned the passing of Christian unity, hoped that all the churches, sects, and quarreling individuals would get together and work for the establishment of peace and justice on earth. But he found little to cheer him. The sermons he listened to so devoutly—how thin and silly, how wrong-headed they sounded beside the sermons preached by his retorts and balances. How much less they told him about the world. Each opinion of each eminent divine could be refuted in just as convincing language in the very next church, while a fact taught by the laboratory might be inconclusive and unimportant, but was certainly based on something better than the dogma of the past.

Boyle was one of the first great scientists of the modern era. He was one of the first to use rigorously the scientific method with its mechanism of observation, hypothesis, and proof. But he didn't do this without regret. He wasn't quite sure that science was good for the world, and the tremendous amount of time he spent apologizing in print for his scientific activities is good proof of this. The scientific method, although advanced by Aristotle, by Roger Bacon, was a new thing in the modern world, and was still under fire as a dangerous rival to religion. Boyle felt this deeply. Never was a man more torn between two spiritual desires. His mind worked scientifically. He couldn't stop it. Everything he observed he added to his stock of knowledge for use in testing the theories it touched. But he was never free from a vague feeling of guilt. What if this monster, science, should finally defeat religion? What if facts should grow until they pushed faith out of the hearts of men? Boyle was uneasy and troubled in his mind. Perhaps he was nourishing a monster which would destroy everything he loved and cherished in the world. But he went on. A born scientist like Boyle can't stop thinking because he fears that his thoughts may be dangerous to humanity.

It was in 1663. The Protectorate had fallen. Charles the Second was on the throne, and Boyle, although known to have Puritan leanings, found himself in a position of great influence. Not only was his reputation as a philosopher at a very high point, but his numerous brothers who'd fought for the king were almost all-powerful at court. Roger Boyle, who had the good fortune to change sides at the psychological moment, was made Earl of Orrery and President of Munster. Richard Boyle became Earl of Burlington and an important creditor of the king. Frank Boyle became Viscount Shannon. And greatest honor of all, his pretty wife, Betty Killigrew, became the king's mistress and the mother of one of his all too numerous children.

Perhaps the moral Boyle looked somewhat askance at this crowning honor. The chances are he did. But he knew that kings from the time of David on had picked the fairest flowers where they grew. It was nothing new and in 1663 Boyle had other worries. The conflict in his soul between faith and scientific skepticism had suddenly come to a head. He could avoid a decision no longer. The Restoration had made him the most powerful and conspicuous natural philosopher at the court of Charles the Second. And Charles the Second had just granted a liberal charter to the Royal Society.

Well, why not? Why shouldn't the "Royal Society of London for Improving Natural Knowledge" have a charter? According to modern ideas it certainly should. But the Seventeenth Century saw the matter differently. Ever since the time of Henry the Eighth, England had been torn by the struggle for domination of three religious groups, the Catholics, the Anglicans, and the Protestants. Many heads had fallen under the ax, and many piles of white ashes had cooled slowly in village squares. But these three groups had differed only in such points as ceremony, doctrines, or church government. No one of the three had shown the slightest tendency to deny the authority of revelation, the inspired nature of the Bible, or the power of God to modify by miracle the natural laws of the universe.

And now came natural philosophy, the wolf in the sheepfold, the cankerworm eating away the heart of faith, the hated new doctrine which proclaimed to the world that it didn't believe anything which couldn't be proved. Quite correctly the three religions read the faint handwriting on the wall. They saw that unless this common enemy was crushed, they would all have to retreat into a gradually shrinking domain of the spirit where science could not pursue them.

Boyle was not the only founder of the Royal Society, which had for many years held meetings under the name of the Invisible College, but he was the most conspicuous among them, and he had used his power at court in obtaining the charter. So it was upon him that the storm of protest broke.

"Do you not realize," wailed the bishops, "that this natural philosophy tends toward the destruction of all religion, true and false? Do you not forsee the inevitable downfall of civilization if the feeble human race should insist on judging for itself instead of following the word of God?"

This might have been expected from the Established Church, but it wasn't only the bishops who thundered. All the conspicuous leaders with semi-theological minds rallied to the attack. Hobbes, who had no love for the clergy, joined in. Butler threw his barbed lampoons. And Henry Stubbs, a copious pamphleteer with a large following, wrote to Boyle most earnestly:

"I beseach you, sir, consider the mischief it hath occasioned in this once flourishing kingdom, and if you have any sense, not only of the glory and religion, but even of the being of your native country, abandon that constitution. It is too much that you contribute to its advancement and repute: the only reparation you can make for that fatal error is to desert it betimes. Do you not apprehend that all the inconveniences that have befallen the land, all the debauchery of the gentry (which ariseth from that pious and prudent breeding, which was and ought to be continued) will be charged to your account? It will be impossible for you to preserve your esteem but by a seasonable relinquishing of these impertinents."

But Boyle was an older man than when he wrote his very pious "Occasional Reflections," and he'd been struggling with scientific problems too long to be turned aside by any amount of righteous denunciation. So he stuck to his guns, and the Royal Society flourished unmolested. It even became fashionable, the merry king in his elegant clothes performing simple experiments with his own hands.

At times Boyle felt certain misgivings. What would be the end of all this? Perhaps he and his friends were sowing a whirlwind which would sweep away all the structure of morality which his pious heart loved so well. But the logical side of his mind wouldn't let him change his course. He still went on, apologizing from time to time in marvelously dull and extended discourses on the usefulness of science, its helpfulness to true religion, and its necessity to the Christian state.

Finally the Church tried bribery. Boyle was offered a bishopric, the provostship of Eton—anything—if he would only be ordained and put the seal of his approval on the doctrines of religion. But Boyle smiled as he wrote a polite and evasive refusal. He was already a bishop, the First Bishop of Science, and his church was the Royal Society.

# The Chemical Revolution

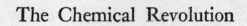

# CHAPTER VI

## The Chemical Revolution

THE scientific method, that stern doctrine, points out a straight and narrow path for the feet of its devotees, and woe to the scientist who turns aside to pick a flower growing off the beaten track. He holds up his bright prize, and the world applauds. But the flower fades at once, and he has to pick another to conceal the fading of the first. Each step takes him farther from the firm ground. The earth grows boggy under his feet; his followers at last desert him, and he sinks to an unhonored grave marked only by a few supercilious lines in the history books of science.

This preacher's allegory of the "straight and narrow path" has been used in every religion to force obedience to various rules from absolute celibacy among the early Christians to the prohibition of buttons among the Mennonites, but to science it applies particularly well. Every scientist is tempted to construct a gaudy, sweeping theory on a basis of too-few observations, and then to ignore subsequent observations which do not fit into the theory. Often the false theory remains standing for a long time, sometimes long enough to attract fanatical devotees. But eventually it falls and the reputation of the erring scientist falls with it.

So it was with Georg Ernst Stahl, the inventor of phlogiston. For nearly a hundred years he was a demigod of science, the inspired prophet of a perfect theory. But phlogiston finally fell. Its enemies celebrated a prolonged feast of triumph, and Stahl, who was a really great man, is remembered only as the author of a false doctrine which misled the world for a hundred years.

Phlogistonism was the Old Régime destroyed by the "chemical revolution," and like the Bourbon monarchy it had roots going back to the Middle Ages. The early alchemists had always been interested in combustion. It was the most spectacular chemical phenomenon they knew. It was easily observed. And it presented a baffling mystery which wasn't cleared up finally until late in the Eighteenth Century. The work of Boyle freed chemistry from the old theories of Elements and Principles. But one of the elements didn't quite die. This was fire, which came to life as Stahl's phlogiston. It was the last gasp of alchemy, but it was a remarkably sustained gasp which lasted nearly a century.

Never did a fundamentally mistaken theory present a more convincing appearance of correctness. It was simple; it was complete; essentially it was this: "Substances which will burn," said Stahl, "are made of phlogiston, the principle of fire, combined with a base. When such a substance burns, the phlogiston leaves the base and combines with the air. When the air has taken up all the phlogiston it will hold, the fire goes out, as in a closed vessel. If more air is added the fire may be started again. Some substances, such as wax, contain instead of a solid base, a gaseous one, which may be detected by bubbling the gas resulting from combustion through lime water and noting the white precipitate."

At first the theory worked perfectly, all observation falling neatly in line in a way to delight the heart of a scientist. It worked not only with the common fuels, but also with metals, which Stahl said contained only a small amount of phlogiston. Stahl heated lead in air. It changed slowly into a red or yellow calx, which according to Stahl was lead deprived of its phlogiston. To turn this back into lead all he had to do was to replace the phlogiston. He mixed the calx with powdered charcoal, which contained much phlogiston, heated the mixture, and sure enough, there was the lead again, bright and metallic as ever.

With true scientific energy Stahl repeated his experiment. It worked every time. It worked with iron, with zinc, with mercury, with sulphur. The theory was perfect, and the late Seventeenth Century celebrated the discovery of phlogiston as one of its major achievements.

There were certain troublesome details, of course, but they quickly retreated before the triumphant advance of Stahl's theory. If, for instance, lead is heated with air in a closed vessel, the air loses some of its volume. "Why is this?" asked the skeptical. "If phlogiston has been added to the air and nothing taken away, why isn't the volume of the air greater instead of less?"

"Phlogiston is not an ordinary substance," replied Stahl. "When it combines with air, the volume of the compound is less than that of the original air." And no one could contradict him.

Another difficulty was that a metal deprived of its phlogiston weighed actually more than before. This caused a momentary flurry in the phlogiston camp, but the attack was safely beaten off, leaving the cherished doctrine more beautiful and symmetrical than ever.

"Phlogiston," said the faithful, echoing the words of Stahl, "is not an ordinary substance, and therefore it does not have ordinary weight. It is the principle of fire, of motion, of levity. It has negative weight. Observe how the air above a fire leaps upward. It has acquired phlogiston, and is therefore lighter. So it rises."

This quieted the critics for a time. The law of conservation of matter was not yet proved, only suspected. Negative weight was not considered impossible. And phlogiston lived on as triumphant as ever.

Toward the end of the Eighteenth Century the phlogiston theory celebrated its seventy-fifth birthday in a hale and hearty condition. Seventy-five is very old for a scientific doctrine, and the phlogiston theory had grown a long white beard and collected a devoted following of orthodox chemists who shouted "Heresy!" whenever their faith was attacked. Phlogiston was entrenched in the textbooks, in the schools, in the minds of the gullible learned who were as numerous then as now. Every new discovery had to agree with it or be denounced as an *ipso facto* error. When hydrogen was isolated, for instance, the phlogistians scored a great triumph. Here was their phlogiston itself, or at least a substance containing a very large amount. It burned with violence; it restored "dephlogisticated" metals to their former state, and most notable of all it was so much lighter than air that it might very likely have actual negative weight.

But at this very moment of triumph, the enemies of phlogiston were gathering. New discoveries were announced on every side. New methods of dealing with gases allowed experiments to yield accurate quantitative results. And a spirit of skepticism was abroad in the scientific world such as had not been seen since Robert Boyle smashed the Peripatetics with the deadly long-windedness of *The Skeptical Chymist.*

The overthrow of the phlogiston theory was called the chemical revolution. It probably got its name from analogy with the French Revolution which happened at the same time, but it was a real revolution nevertheless. It jolted men's minds out of their comfortable grooves. It destroyed a cherished doctrine which had been taught for decades, and it laid down the basic principles on which chemical research has proceeded from then to the present time.

The credit for the chemical revolution goes chiefly to three men, two Englishmen and one Frenchman— Joseph Priestley, Henry Cavendish, and Antoine Laurent Lavoisier. Each one played his part, the Englishmen contributing the most important discoveries, and the Frenchman, after the manner of Frenchmen, putting them together in a complete and orderly theory. In this chapter we'll consider the discoveries themselves, and show how each was a step toward the revolution. Then in later chapters we'll deal with the three chemists as men. They were very different, these three, and they arrived at their conclusions by very different methods. But each one is a fascinating study of genius.

Up to the middle of the Eighteenth Century science didn't know much about gases. They were all thought to be modifications of air, or mixtures of air with various vapors, spirits, or fumes. It was this ignorance which allowed the phlogiston theory to live so long. If Boyle had carried his scientific work a little farther, the phlogiston theory would never have been born. Gases are the most docile and amenable of chemical substances. They are limited in number. They react quickly and form comparatively simple compounds. But unless the chemist has special apparatus to handle them, they are elusive and inconspicuous, difficult to distinguish from the innumerable bad smells of the laboratory. A key invention was needed which would allow chemists to deal with gases effectively.

The key invention was made by Priestley. It was called the "pneumatic trough," and was so simple and perfect that there was probably a great chorus of "Why didn't *I* think of that." The pneumatic trough is nothing but a large vessel containing water a few inches deep. Priestley used an earthen washtub. A

shelf with several holes is placed just below the surface. Wide-mouthed bottles filled with water are inverted over the holes, and tubes are passed upward
into them. The gas rises into the bottles, replacing the
water. When one bottle is full, it may be slid aside,
and another substituted. Never did so simple a device fill a crying need more perfectly. There was no
danger of loss of gas or contamination with anything but water vapor, and this could be avoided
when desirable by using mercury instead of water.

Equipped with this homely device, Priestley got
to work, starting with carbon dioxide, or "fixed air,"
which he obtained from a near-by brewery. He didn't
discover much of scientific value. Carbon dioxide was
to remain something of a mystery for many years.
But he did observe that the gas dissolved in water
had a "pleasant acidulous taste" and he developed a
method of making "an exceedingly pleasant spar-
kling water which could hardly be distinguished
from very good Pyrmont or rather Seltzer water."
He passed the hint to a commercially minded friend
who bottled the first soda water, which he called "the
Mephitic Julep." Thus Priestley, besides being one
of the several "fathers of chemistry," is the unchallenged patron saint of the soda fountain. Which may
or may not be a credit to him.

The next gas Priestley tackled was "inflammable

air," or hydrogen. This had been known for a long time, for the most addle-headed alchemist could hardly help observing it bubble off when a metal was dissolved in weak acid, but it had never been investigated thoroughly. Most chemists who knew anything at all about inflammable air thought it was phlogiston. Priestley thought so too, and his first experiment was intended to prove it.

He took a quantity of "dephlogisticated lead," which we call lead oxide, placed it at the top of a tall glass cylinder inverted over water and filled with inflammable air. Then he waited for a sunny day and heated the oxide with a large burning glass. It became black, turned into metallic lead, and the inflammable air disappeared at a great rate, drawing the water up into the cylinder. Priestley rejoiced. He thought he had certainly proved that inflammable air was phlogiston for it restored dephlogisticated lead to its metallic state without leaving any residue. Of course we know now that water was formed in the cylinder. But Priestley had no way of observing this, for the cylinder was full of water already. Even when he used mercury, the amount of water was so small that it might easily have come from the imperfectly dried hydrogen or from the oxide itself.

Priestley repeated this experiment on various

metallic oxides with the same result. The inflammable air reduced them all. Nothing else appeared to be formed in the process. He felt sure he'd found the genuine phlogiston.

By this time it was 1774 and Priestley was about to make his most important discovery. Encouraged by the success he had with the combination of pneumatic trough and burning glass, he set to work trying to extract an "air" from every substance he could lay his hands on. He isolated sulphur dioxide, ammonia, silicon fluoride among others. But his crowning triumph was the red oxide of mercury.

He placed some of this in a glass bottle, filled the bottle with mercury, and turned it upside down in a vessel containing more mercury. The oxide rose to the top, and Priestley heated it through the glass with his invaluable lens. An air was given off which soon filled the bottle.

Priestley transferred his new air to a cylinder standing over water and found that it was not "imbibed" as the ammonia and silicon fluoride had been. And on trying to ignite it with a candle he found that not only did it refuse to burn, but the candle itself flared up with tremendous brilliance and burned with a white flame until all the air was exhausted. Priestley, needless to say, was astonished and delighted.

Nothing like this had ever been seen before. He jumped to the conclusion that this air was completely dephlogisticated, contained no phlogiston at all, and so allowed the phlogiston in the candle to escape more rapidly. Priestley had made perhaps the greatest single discovery in chemistry, but he never realized its magnitude. His obstinate conviction that his discovery was not a new gas, merely "dephlogisticated air," prevented him from any really valuable conclusions. This was left for Lavoisier.

As early as 1770 Lavoisier had been convinced that the phlogiston theory was a delusion. For one thing, he had firm faith in the law of conservation of matter, and he considered the negative weight of phlogiston ridiculous. All matter, said Lavoisier, weighed something at least. "But phlogiston," said the faithful, "is not ordinary matter. It is matter *transmuted* into something else. And if you don't believe that matter can be *transmuted,* evaporate the purest water and observe the 'earth' left behind. Water has been *transmuted* into earth. And this proves that some things in the world are beyond the power of your theory to explain."

The reasoning here is a bit muddy, but the phlogistians had raised a very real objection to the theory that a chemical element can enter into combinations

with others, but cannot change into another element
or vary in weight. Lavoisier saw that all the reasoning he intended to use to destroy phlogiston would
fall to the ground unless he removed this objection.
If he couldn't prove that the elements were indestructible, he couldn't prove anything by the careful
quantitative demonstrations he had in mind. His
opponents would always be able to take refuge in
vague references to "transmutation" and "other
kinds of matter."

The delusion that distilled water was transmuted
into earth had remained unchallenged up to this
time because no one had been able to weigh with sufficient accuracy the glass vessel in which the water
was evaporated. The amount of earth left behind by
the water was so exceedingly small that the large
scales necessary to weigh the vessel could not detect
any loss of weight. Lavoisier reasoned that if, instead of merely evaporating the water, he should heat
it for a long time in the same vessel, enough glass
would be dissolved to register on his scales.

Lavoisier was certainly patient. He heated the
water nearly to boiling for no less than one hundred
and one days, and when he finally evaporated it he
obtained 20.4 grains of earth—more than a gram,
and enough for any good balance to deal with accurately. Then he weighed the vessel and found that

it had lost 17.4 grains, which was near enough to prove his point. The water was not transmuted into earth. It merely dissolved a small amount of the glass from the vessel. Lavoisier had demolished the chief objections to the law that the elements are indestructible. Now he could go ahead with firm ground under his feet.

The first thing he did was to repeat the celebrated experiment by which Stahl proved that charcoal "restored phlogiston" to "dephlogisticated" metals. He heated lead oxide and charcoal together in a cylinder on Priestley's pneumatic trough and found that, instead of merely exchanging phlogiston, the two substances gave off a gas which he identified as the fixed air which Priestley got from the brewery. Here was something new for the phlogistians to explain.

The next step was a crucial one. The phlogiston theory maintained that when a metal burned it gave off phlogiston of negative weight and thus became heavier. Lavoisier now proceeded to demonstrate that what really happened was that the metal combined with part of the air and thus gained as much weight as the air lost. To do this he had to carry out the whole experiment in a closed system or he wouldn't be able to weigh the air after the metal had combined with part of it. The apparatus he used is so simple and ingenious that it's worth describing. It is still used whenever quantitative work has to be done on gases.

Lavoisier put a weighed amount of granulated tin in a glass retort with a narrow tubular opening, drew out the tube to a thin hair and sealed it with a blowpipe. He then heated the tin until it had formed as much oxide as it would. No air or anything else had entered the retort, and therefore the whole apparatus weighed as much as before. Lavoisier let the retort cool, and broke off the glass filament. The air rushed in with a hissing sound to replace the part absorbed by the tin. The weight of this additional air was found by weighing again the retort and its contents. Lavoisier then weighed the tin and tin oxide and found that they had gained exactly as much as the whole apparatus had gained when the air was allowed to enter. He had proved that when metals are burned they do not lose phlogiston, but absorb from the air a substance which increases their weight by exactly as much as the air loses.

This was exceeding interesting; it confirmed Lavoisier's disbelief in phlogiston. But it wasn't final. He didn't know what it was the tin had taken from the air. He needed another fact before he could offer a complete explanation. By a fortunate accident this fact was given him at just the right time.

In 1774, while Lavoisier was engaged on this prob-

lem, Priestley visited Paris and was entertained by
Lavoisier at a dinner "with most of the philosophical
people of the city." He mentioned his new dephlogisticated air, in which a candle burned so brilliantly,
and explained how he got it from mercuric oxide.
Lavoisier "expressed great interest," and well he
might, for here was the information he needed. No
doubt he waited impatiently for the next morning
when he could go to his laboratory and make some
dephlogisticated air of his own.

Now mercuric oxide is peculiar stuff. If you heat
mercury nearly to its boiling point in air, the oxide
forms slowly on its surface, but if you heat it any
hotter, the oxide breaks down, forming mercury and
oxygen again. It is what chemists call a "reversible
reaction." Priestley may have known this, but he
made no use of it. Lavoisier did. He heated mercury
nearly to boiling for twelve days in a closed retort.
The mercury absorbed eight cubic inches of air, and
a small amount of oxide formed on its surface.
Lavoisier collected the oxide, heated it strongly, and
expelled from it eight cubic inches of dephlogisticated air, which behaved exactly as Priestley said it
would and proved that the mysterious gas was nothing else than the constituent of air which combined
with metals to form oxides.

Lavoisier was so encouraged by this that he rather

lost his head and began to generalize a bit too soon. When he found that the new gas would combine with carbon, sulphur, or phosphorus to form substances whose water solutions were acid, he jumped to the conclusion that dephlogisticated air was the "acid-forming element." So he coined the name "oxygen," which means "acid former" in Greek, and the name has stuck ever since, although it should have belonged to hydrogen instead.

This discovery was a pretty bad blow to phlogiston, but still it wasn't conclusive. There was one more thing to explain. "This is all very interesting," said the faithful, "and perhaps true. But why does inflammable air restore dephlogisticated metals to their metallic state?" Lavoisier couldn't answer. He was stumped again. He needed another new fact from England.

For the present let's leave Lavoisier rather discomforted by this counter-attack and consider Cavendish, who was about to find the fact which, like the key-piece of a Chinese puzzle, would make all the others fit together into a symmetrical pattern.

Cavendish had been investigating off and on for some time the properties of hydrogen, or inflammable air. At the suggestion of Priestley he decided to explode a mixture of dephlogisticated air and inflammable air in a closed vessel so that he could capture whatever was formed. If inflammable air was phlogiston, as he thought, and dephlogisticated air was air deprived of its phlogiston, as Priestley thought, a combination of the two in the proper proportion ought to give nothing but ordinary air with its normal amount of phlogiston. With great technical skill Cavendish exploded various mixtures of the gases. Never did he find anything left but a surplus of one or the other. And when he used two parts of hydrogen to one of oxygen, the water over which his vessel stood rose actually to the top, occupying nearly all the space where the gases had been. It was easy to see that no ordinary air was formed.

Cavendish was baffled for a while, but finally he remembered that several years before Priestley had demonstrated to James Watt Stahl's famous experiment of using inflammable air to reduce lead oxide, and that Watt had noticed a small amount of water in the vessel. Neither Watt nor Priestley attached much importance to this at the time, agreeing that the water probably came from the imperfectly dried hydrogen. But now Cavendish saw the light. He exploded another mixture of hydrogen and oxygen, over mercury this time instead of water, and found that a quantity of water was formed whose weight was exactly that

of the two gases. Which proved that water, instead
of being the typical simple substance, as all earlier
chemists had thought, was a compound with a definite formula which we now write $H_2O$.

When Lavoisier heard about this, he ran to his
laboratory with what would have been, had he not
been an atheist, a thankful prayer to the Lord. Here
was the last bit of information he needed to complete
his theory. Now everything was clear. Lavoisier
made a few confirmatory experiments, and rushed
into print. Phlogiston had finally fallen.

Lavoisier's theory of combustion, on which all
modern chemical reasoning is based, was simple and
complete. It can be stated in a few lines. The air
consists of two gases. One of these, oxygen, combines
with metals and other combustible substances to form
oxides. When it combines with hydrogen it forms
oxide of hydrogen, or water. Hydrogen can take the
oxygen from a metallic oxide, forming water, and
leaving the metal behind. That was all. No more was
needed. The litter of phlogistonism was cleared
away, and chemistry could build up firm theories on
firm foundations.

Such was the chemical revolution, and seldom has
a change of scientific opinion had as far-reaching
consequences. For the first time the chemists touched
bottom, penetrated the labyrinth of conflicting

theories and observations, and captured the few simple but potent truths which were all they needed. From here on the progress of the science was rapid and continual, without a single backward step. Old mysteries were cleared up almost daily. New discoveries came in rapid succession. And modern scientific civilization, which is founded on chemistry, could celebrate its first birthday.

Before ending this chapter let's say a few kind words for Stahl and his phlogiston theory. It caused a tremendous amount of confusion and did a great deal of harm, but, looked at broadly, it wasn't so far from the truth after all. Stahl merely saw things in reverse. If instead of saying that burning substances *give off* phlogiston, he had said that they *absorb* it from the air, he would have been as nearly right as could have been expected. Oxygen takes part in all ordinary combustions, and to call it phlogiston, the "fire substance," isn't very far off.

The trouble with the phlogiston theory was that it became a dogma of the theological-minded. It inspired too much blind devotion. Men were determined to stick to it in the face of contradicting facts, and were willing to go to almost any extremes rather than give it up. The clear and rational start which Boyle had made with his *Skeptical Chymist* was forgotten completely when science was hypno-

tized by Stahl's seductive theory. It took the vision
of Lavoisier and the revolutionary discoveries of
Priestley and Cavendish to penetrate the smoke
screen raised by the followers of Stahl around the
fundamental problem of combustion.

# Joseph Priestley, the Minister Who Wanted To Believe

# CHAPTER VII

## Joseph Priestley, the Minister Who Wanted To Believe

THE Dissenting congregation of Needham Market had taken a new minister on trial, and this was his first Sunday service. No one had seen him yet except a few of the elders, and curiosity was intense. The chapel was a small and poor one, but it was full to the doors. The arrival of a new minister meant a lot in 1755. Even the stay-at-homes were there.

They couldn't hope for much, thought the congregation gloomily. All they offered was forty pounds a year, hardly enough for a farmhand, let alone a minister. He'd have to live on two shillings a day and rent a house for himself into the bargain. So with more apprehension than hope they watched the little door behind the pulpit.

"He's only twenty-two years old," they thought, "and that's in his favor. An experienced minister for forty pounds a year would be a strange creature indeed."

The door opened and the new minister took his place in the pulpit. He was rather tall, rather good-looking, with an open gentle face and a pleasant smile. He made a very good impression in the first few moments while he looked through the Bible for his text. But when he began to speak a frown of annoyance spread over the faces of his listeners. He stammered. He stammered painfully. He'd say a few words in a soft musical voice, then stop dead while his jaw moved rapidly and a strangling sound came from his throat.

The congregation squirmed nervously until the spasm was over and the soft voice was flowing smoothly again. There'd be a few fine sentences, as fine as they'd ever heard. Then the gasping silence would begin again, and they'd fidget on their benches and pray for the nice young man who was having such a terrible time up there in the pulpit.

The sermon went on, however, for the new minister had courage and determination. The Lord had afflicted him, for some good reason no doubt, but they all could see what a fine young man he was. After the service was over, the congregation broke up into little groups to discuss their new spiritual leader.

"His name's Priestley, is it? That's a poor name for a Presbyterian, but it's no fault of his. The stammer is terrible, but the words are good when they come. He's a learned young man all right."

"Is he sound on the Trinity?"

"That we don't know"—this from an elder. "He talks above our heads entirely. All Latin and Greek. But he's learned, and the more a man knows the sounder he is on the Trinity. Only the ignorant deny the truth."

"And what a nice friendly smile he's got"—this from a woman. "Perhaps he'll get over his stammer. Perhaps he was nervous because of his first service."

The elders had a meeting that afternoon and voted to retain Priestley in spite of his stammer. Perhaps they were influenced by other matters than his qualifications. Every previous minister had demanded the full forty pounds, ten of which came from a handful of stiff-necked Independents whose presence in the congregation made nothing but trouble. But Priestley offered on his own initiative to do without the ten pounds for the sake of getting rid of these trouble makers. The Independents with their noisy sternness were forthwith ejected, and the little congregation felt that this alone was worth many a stammer.

During the three years that followed, Priest' lived a painfully isolated life. His congregation was small, poor, and ignorant, mostly tradespeople and workmen. He had few points of contact with them. He didn't understand them, nor they him. But they let him rather completely alone on matters of doctrine, and he let them alone on matters of money and church government. He respected their ability to handle their own affairs, and they respected his learning. It wasn't every congregation that could get such a minister.

"He's got a book in his study a foot thick," said an elder in an awed voice, "and it's printed with letters like nothing I've seen before."

"That's a Hebrew concordance, so he told me. It cost him three guineas, so he said. And the Lord only knows where he got the money. It's a month's salary. But if he wants to put his pay into books instead of victuals, it's none of our business. We get the benefit of the books through his reading them. We'd get nothing from the victuals."

So passed the first three years of Priestley's life as a minister. He very nearly starved, and he very nearly died of loneliness. And he was tortured by a guilty suspicion that if these people knew what he really thought, they'd feel that the three years of his ministry had brought them a good deal too near the fires of hell. Perhaps he should tell them? Perhaps he was living a lie? But no. He wasn't sure yet in his own mind. Hidden among the crabbed

letters of his Hebrew, the graceful fronds of his
Syriac, the pleasant music of his Greek, he had
found things which made him doubt. But he didn't
know the whole truth yet. Perhaps the roots of the
faith were buried deeper still. Until he'd found out
he'd keep silent and try to preach to these good
people in words they could understand. They paid
him as much as they could afford; they stood for
his stammering; they admired his learning. They
deserved the best he could give.

Such was Joseph Priestley in his early twenties,
the stammering minister of a feeble congregation.
Totally unknown, totally without resources; earn-
ing the wages of an agricultural laborer and nearly
starving in an atmosphere of Hebrew, Syriac, and
Greek. Let's see what brought him to this position
of learned but painful poverty.

★    ★    ★

The house of the Priestleys at Fieldhead in Yorkshire was long and low and picturesque. It stood on
four distinct levels, its gables climbing uphill like
a flight of stairs. It was surrounded with a wall covered with vines. You'd never have taken it for a
factory, but a factory it was. For this was before the
industrial revolution, and Jonas Priestley was a
weaver.

Nowadays, when we think of a weaver, we visualize a puny, half-starved creature who stands in the midst of a score of domineering machines and listens fearfully for the sound which tells him that something has gone wrong. He gets paid just enough to keep his feeble body and feebler soul together, and if he's wise he checks his individuality at the door on the day he gets his job. His function in the factory is something like the function of the duodenum in the body. He supplies raw material for the machines to work on, and he eliminates waste products such as snarls and broken threads. His character resembles that of the duodenum—quiet and not too full of hope.

But in the early Eighteenth Century things were very different. The weavers were independent manufacturers on a small scale, and very much in charge of their own affairs. They acknowledged no superior; they received no wages. They owned their home, their workroom, and their looms. In short, they were just the "sturdy yeomen" about whom democrats sentimentalize so fervently to-day without noticing that the class disappeared a hundred years ago.

The Priestleys were weavers and cloth finishers rather more prosperous than most, and they thought very well of themselves. In the previous generation some of the family had gone into trade and become rather wealthy, and the fact that there was no barrier of occupation between the two branches of the family shows the substantial position of the weavers in their small communities.

Many of the weavers, and the Priestleys among them, were Calvinists, for the north of England, although it had supported the king in the civil wars a hundred years ago, had become a stronghold of the most rock-like variety of Presbyterianism. No Puritan divine of early New England could excell the Dissenting ministers of Yorkshire in theological frightfulness, and to most of the members of their congregations, the flames of hell were as real and near at hand as the fire on the kitchen hearth.

It was into such a household that Joseph Priestley was born in the year 1733. He was the eldest son, and five more children followed in the rapidest possible succession. Since this increase in the family put a strain on the accommodations at home, he was sent to live with a grandfather who had more room. When he was six years old his mother died in childbirth, and the family was broken up. Joseph was considered the most promising boy, and his father's sister, Mrs. Sarah Keighley, offered to take full charge of him and see that he got a good education.

The offer was accepted at once. Joseph left his father's home for good.

This change is what started him on an intellectual career. Otherwise he would probably have become a weaver like his father, for although Jonas Priestley was "God-fearing," "sturdy," and all that, there's no reason to think that his mental horizon was any wider than his weaving and the harsh doctrines preached in the local chapel. But with Aunt Sarah things were very different. Her husband was a man of considerable wealth, and her home at Leeds was the meeting place of all the ministers in the surrounding country. She was a Calvinist, or called herself that, but she was broadminded within certain limits, and she liked to hear all shades of religious opinion expressed freely around her hospitable tea table. When Joseph Priestley was twelve years old he was allowed to sit in a corner while the ministers talked. Children of this age have usually very little interest in the doctrines of infant damnation and original sin, but Joseph looked forward to these discussions with tremendous interest. He said nothing. Self-conscious children who stammer know better than to talk. But he listened carefully, and if he didn't understand a tenth of what he heard, he did master very completely the principle that in matters of religion only the least intelligent men are orthodox on all points.

And if a child's learned this, he's learned something most adults never do.

All this was wonderfully fortunate for Priestley, the future scientist, and wonderfully fortunate for the world. As we shall see later, Priestley had an orthodox streak in his character. *He wanted to believe* the opinions of the past. If he'd been trained in early youth to accept at its face value every rigid doctrine accepted by the people around him, this trait might have dominated his character, and the world would have gained one more orthodox divine and lost a fruitful scientist. But the atmosphere of his aunt's household was liberal, and Priestley grew up to think for himself and investigate with open eyes the shaky doctrines which the orthodox professed to be so sure about.

His schooling was informal and irregular. His health, which was never very good, did not allow him to follow any rigid course. But he was naturally studious, and he learned by himself far more than most boys learned under the strictest discipline. When he was nineteen he entered Daventry Theological Academy to prepare for his natural profession, the ministry.

For a theological academy, Daventry was a rather unusual place. The spirit of inquiry was fostered, and the students were urged to read as much as possible, whether the books they read agreed with current Presbyterian doctrine or not. Discussion was encouraged, and every opinion advanced by a student, no matter how radical, was considered worthy of a serious hearing. There was no censorship of ideas, no flustered covering up of doctrinal flaws. The slogan, so often proclaimed but seldom applied, that "truth will prevail" was the actual motto of poor obscure little Daventry Academy.

When Priestley became famous, when his name was known from one end of Europe to the other, he had the pleasure of boasting about his humble Dissenting college to the Oxford-bred Prime Minister of England himself:

"Shutting the doors of the universities against us, and keeping the means of learning to yourselves, you may think to keep us in ignorance and so less capable to give you disturbance. But though ignominiously and unjustly excluded from the seats of learning, and driven to the expedient of providing at a great expense for scientific education among ourselves, we have had this advantage, that our institutions, being formed in a more enlightened age, are more liberal and therefore better calculated to answer the purpose of a truly liberal education. Thus while your universities resemble pools of stagnant water secured by dams and mounds, ours are like rivers, which, taking their natural course, fertilize a whole country."

★  ★  ★

After his graduation from Daventry Priestley filled various pulpits and taught at Warrington Academy near Manchester, but everywhere he was pursued by the same problem. He knew too much. His mind was too absorbent for the ministry. He wanted to believe. He tried hard to believe, but he also tried hard to find out the truth, and although these two activities are proclaimed constantly to be identical, they are decidedly not—as any minister will tell you if you gain his confidence. Priestley read many languages—all languages in fact which had any application to theology or natural philosophy. He knew the Bible almost by heart, and he formed the habit of looking through it for the roots from which the various doctrines of Christianity had grown. Often he found none, or insufficient ones. Still hopeful, he'd resort to the difficult original languages—the Hebrew, the Greek—and hope to find clear proof among their twisted letters. But no. He got no confirmation here either, and he had to admit he was puzzled—painfully puzzled.

At this point it might be remarked that Priestley was a bit naïve. He *was* naïve—very naïve. Never did

a more trusting soul attack a more untrustworthy subject. His mind at this period was working like that of
a scientist. He actually thought that the doctrines of
religion could be tested as you can test the belief
that the square of the hypotenuse equals the sum of
the squares of the other two sides. He didn't realize
that scientific reasoning and religious reasoning are
fundamentally different; that while a scientific belief
is only as strong as the facts which support it, a
religious belief needs no such roots. It is like an
orchid which sprouts in the crotch of a tree and
develops roots later if at all.

After Don Quixote had attacked the windmills
so valiantly, it is likely, although it is not recorded,
that he soon found something more substantial to
attack—the owner of the windmills, for instance.
So it was with Priestley. The type of reasoning he
had been applying to religion was purely scientific,
and he naturally got no more results than Don
Quixote got with his lance. Soon he turned with a
mighty sigh of relief to problems which were solid
enough to get a grip on, problems which would
come out in the open and be defeated. In this indirect way Priestley approached natural philosophy.
He discovered many things for which the world
should be, and is, duly grateful. But he never quite
abandoned theology. At the bottom of his heart he

always felt that science was a rather empty, shelly thing, not nearly so essential to human happiness as religion with its great problems of faith and morality. We shall see how this fundamental theological-mindedness affected to the day of his death the nature of all his scientific work.

Priestley's first activity in science was very humble indeed. While at Needham Market he tried to eke out his meager salary by giving twelve lectures on "The Use of a New and Correct Globe of the Earth." He had only ten listeners, and the whole proceeds hardly paid for the globes. But Priestley wasn't discouraged. He had the globes, which was what he really wanted; he'd had a good time talking about them; and he wasn't starved yet. When he moved to his next church at Nantwich in Chester, he opened a school which was an immediate success. For the first time he had enough money to buy the "philosophical instruments" he longed for—a small air pump and an electrical machine which he trained his pupils to operate and keep in order. Priestley was an excellent teacher, and the parents of his pupils were very completely satisfied with his work. But the pupil that learned most was Priestley himself. He had apparatus; he had leisure; he had plenty of willing boys to help him. And when in three years he moved on to teach at Warrington Academy, he

was well started on the road toward scientific knowledge.

Priestley's activities at Warrington give an idea
of the scope of his interests. He wrote and lectured
on every subject which retained his attention for
even a moment, and his writings found a ready
market in that pamphlet-consuming age. He was a
born pamphleteer, never rewriting, never composing
painfully, but setting down the thoughts as they came
to him, and setting them down in final form ready for
the printer. He wrote on biography, on education,
on economics, on a dozen different aspects of theology. At this period he published nothing on
science. He didn't know enough for even a pamphlet.
But the turning point came during a visit to London
in 1766, for it was then that he met Benjamin
Franklin.

In 1766 Franklin was one of the most conspicuous
figures in London. He and his friends in Parliament
were fighting the great battle for the American
Colonies which resulted in the repeal of the Stamp
Act. He was courted by statesmen and fought over
by society, but somehow he found time for more
purely intellectual matters. And when he met Priestley at the home of a "philosophical friend," he
recognized in the serious, stammering young minister
the great talents which as yet no one else had noticed.

J. PRIESTLEY, LL.D. F.R.S.

JOSEPH PRIESTLEY
The Unitarian preacher who fled from English mobs
to America and became the first great Americanized
chemist.

It must have been a strange friendship which developed between Franklin and Priestley—between the perfect self-made man of the world and the shy young scholar from the land of cloth and Calvinism. Franklin was sixty years old then and Priestley only a little more than thirty. Franklin was a wise old diplomat who'd risen to international importance from the greasy obscurity of a Boston tallow shop. Priestley was a scholarly recluse whose life was abstract thought and speculation alone and who had no interest in political or social prestige. But their friendship developed nevertheless and lasted in its intimacy all through the American Revolution to the year of Franklin's death in 1790.

Now Franklin had various talents, some of which patriotic Americans pretend to forget, but his chief talent was for the judging of men. When he met Priestley he sized him up at once as a man of first-rate ability whose energies were being wasted in nonproductive fields. Theology, English grammar, elocution! These, thought Franklin, were sorry subjects for a great mind to work on. Leave them to small minds, the smaller the better, for when there's nothing to discover, it's a shame for a good mind to waste itself in searching.

Franklin was strong on good advice. He probably gave more good advice than any man in history, and he knew how to adapt his advice to the needs of the advisee. His greatest triumph in this respect was when he urged Priestley to write a history of electricity and offered to supply him with the necessary books and information. Never was a better suggestion given at a more opportune time. Priestley knew very little about natural science, but he had the intellectual equipment to learn. Compiling a history of electricity was just the task to get him started. He'd have to learn what had been done by others; he'd have to test their experiments with actual apparatus, and if he made a good job of the book he'd win an ample reward in praise and publicity—trust Franklin and the Royal Society for that. Almost all the available information about electricity was contained in the *Philosophical Transactions* of the society, and even in the Eighteenth Century scientific institutions were duly grateful when their publications were quoted as the final authorities.

Everything came out exactly as Franklin had hoped. Priestley went back to Lancashire fired with a new and burning enthusiasm and loaded down with copies of the *Philosophical Transactions*. In a little more than a year the history was finished and was a great success. Its publication gained Priestley immediate recognition as a man of science and secured his election to the Royal Society.

About this time he left Warrington Academy and moved to Leeds to take charge of a liberal congregation which enjoyed having a famous natural philosopher for its minister. It was perhaps the freedom which his tolerant church allowed him, or perhaps a desire for recreation after his scientific labors, but at any rate Priestley now indulged in a prolonged orgy of theological writing. The presses had a hard time keeping up with his agile pen. He wrote on every imaginable subject from "Family Prayer" to "The Institutes of Natural and Revealed Religion." And before he'd been at it long, he found himself at the storm center of the most violent sort of religious controversy.

There's no use going deeply into Priestley's theological opinions. No subject is more obscure and meaningless to the modern mind. But one divergence from the orthodox faith is worth mentioning, for it caused most of the trouble that was coming to Priestley and had a profound effect on his later life. He gradually became a Unitarian, or, as it was more commonly called then, a Humanitarian. He believed that Jesus was a man, inspired perhaps, but not a God or a son of God. This seems a harmless thing to believe, but in that day nothing could have been more radical or dangerous. There's a twist in the religious mind which makes a heretic much more hateful than an unbeliever. The intellectual leaders of the time were atheists or freethinkers almost to a man, and the faithful seemed reconciled to the situation. But when Priestley became a Unitarian, he brought down on his head all the concentrated wrath of the godly. Even Gibbon, that arch enemy of Christianity, denounced him in no uncertain terms, apparently because he thought a minister of religion should stick to the fundamental doctrines or leave the Church entirely.

The chorus of abuse which assailed Priestley's Unitarian opinions would have been a fine show for a man who could enjoy it. He was denounced in mellow tones from the dim aisles of the Established Church, in shrill screams from the chapels of the Dissenters, and in the deep bass of the natural philosophers who thought with Gibbon that ministers of religion should stay strictly on the reservation. But Priestley could not enjoy it. He was deeply hurt at the reception of his doctrine. He had a mind which made him investigate and doubt, but deep within his heart he was like Robert Boyle. He *wanted to believe*. We'll see presently how he found a faith, and how sadly that faith betrayed him.

Up to this time Priestley had done little work in chemistry. He knew almost nothing about the science, and consequently when he started to learn he turned

to the published works of the past. This is all right
in theory. It's just as well to know what has been
done before you start work yourself. But in this case
it was very unfortunate, for the textbooks of the time
were full of the fair but false phlogiston theory,
and Priestley was hypnotized by its misleading appearance of perfection. "Here," he said, "is the
truth. I've searched through the Scriptures and
found nothing but confusion. I've examined the
science of electricity and found nothing but isolated
and unconnected facts. But here is a firm founda-
tion. Here is a solid something to start with."

Then and there Priestley dedicated his scientific
life to the phlogiston theory and stuck to it through
thick and thin. He lavished on it all the devotion
which his skeptical mind would not let him give to
any of the hard-and-fast systems of religious dogma.
And what a poor subject of devotion it was! Priestley was like a very uninspired rat running up the
hawser of a ship about to set sail for Davy Jones's
locker.

The results of this bad choice of scientific faith
did not appear at once. For a time all was serene.
The phlogiston theory was wrong, but it wasn't a
bad leader to follow for a short distance. It wasn't
due for complete destruction until Priestley himself
had discovered the facts which finally destroyed it.

Working, as he thought, to prove the ultimate truth of the phlogiston theory, Priestley unwittingly became one of the three fathers of the chemical revolution which forced it to the scientific junk heap.

When Priestley began his chemical work he was living at Leeds, and the building next door was the public brewery of Jakes and Nell. This institution was appreciated by all the neighbors, but by Priestley doubly so, for not only did he use its finished product on occasion, but the carbon dioxide given off by the fermenting beer was the subject of his first chemical research. This gas, then known as "fixed" or "mephitic" air, had been investigated more or less, but only imperfectly. The best authorities, such as they were, thought it was "phlogisticated air" or air which had taken up as much phlogiston as it could and which therefore would not support further combustion. Priestley held to this opinion to the end of his life although the very discoveries which he made himself had proved it incorrect.

At this period he knew very little about chemical apparatus, and he had no money to buy the apparatus he did know about. Far from being a handicap, this proved a tremendous advantage for it made him develop ingenious methods of his own which turned out far more accurate than any used by his predecessors. He was forced, for economy's sake, to use the objects he could pick up around the house, and a large earthen washtub was the first thing he selected. Next he made a wooden shelf which fitted into the tub and stood a few inches off the bottom on cleats. In this he bored a number of holes. A few glass jars from the pantry and a little glass tubing completed the device. So far it had cost almost nothing. The only part which had to be paid for was the tubing, and since this was made merely by pulling out a glass bulb, it was nearly as cheap then as now. No doubt the outfit looked rather crude when Priestley put it together, but to the present day no real improvements have been made upon it. We don't use washtubs or preserve jars, but we could if we wanted to, and we'd get just as good results.

Laden with this curious apparatus, Priestley proceeded to the brewery and was welcomed by the brewer, who was probably glad of a chance to get on good terms with the clergy. He set his tub on a barrel near one of the fermenting vats, filled it and the jars with water, and inverted the jars over the holes in the shelf. Then he bent a length of the glass tubing over a spirit lamp so that one end would stick up into the mouths of the jars and the other rest on the rim of the vat just above the bubbling beer. He took his finger from the upper end of the tube. The gas rushed into the jar, replacing the water, and

Priestley had the first pure sample of carbon dioxide.

This doesn't sound like much of a scientific experiment, and in one sense it was not. But it illustrates how Priestley worked, how he used common sense to solve problems which had baffled men more skillful and experienced than he. All previous attempts to isolate a pure gas had failed, for some air had always got into the bulbs and globes of Priestley's predecessors. He took his gas from a pure source, the covered vat, and allowed it to touch nothing but water, which dissolved some of it, but left the rest uncontaminated.

Priestley filled all his jars with the gas, sealed them carefully, and bore them back in triumph to the shed he used as a laboratory. At this stage he didn't know what it was he had in the jars. In fact he never did know exactly, for the phlogiston theory always managed to get between him and the truth. But he had the gas safely in his jars, and he set to work to learn its properties.

It must be admitted that he didn't learn much at this time. He was too much of a beginner. But he did make one discovery which, although hardly scientific, did win him great renown. This was the invention of soda water. He noticed that the water over which fixed air had stood absorbed a certain amount of the gas and acquired a "pleasant acidulous taste." For some obscure reason he thought the solution would cure scurvy. The Admiralty, to which the matter was brought, thought so too, and Priestley's apparatus was installed on two of His Majesty's battleships. So when Englishmen wax scornful about soda fountains, the patriotic American may point out that the first two were set up by the British Navy itself.

The invention of the pneumatic trough and the discovery of soda water won Priestley great recognition, and the Royal Society gave him the Copley Medal, the highest honor in its power to bestow. Nor was this all. Lord Shelbourne, late Secretary of State under the elder Pitt, offered him a position as "literary companion" with a salary of two hundred and fifty pounds a year—two and a half times what he was getting at Leeds. Priestley accepted after some hesitation, and for the first time in his life was free of all financial worries and able to buy the equipment to carry on his research.

While Priestley was living with Lord Shelbourne at Calne he did the greater part of his productive work. His duties as literary companion were very light indeed, and Lord Shelbourne saw to it that he had everything he needed. It was here that he discovered oxygen and the other gases so important to chemical theory, and it was here that he wrote the books which have made him immortal as a chemist.

We have explained in the preceding chapter the significance of Priestley's discoveries. Perhaps no single man has made more important observations. His notebooks were a mine of scientific hints for many years. But the most striking thing about Priestley was that he never realized the full importance of his work. He discovered oxygen, but he never admitted that oxygen existed, maintaining to the last that it was only air deprived of its phlogiston. He was the first to observe the formation of water from its constituents, but he attached no importance to it. He made many other observations of the utmost importance, but never drew from them correct conclusions. While experimenting with carbon dioxide he tried growing some mint in a jar of the gas. The mint grew marvelously well, and Priestley was amazed to find that the gas in which it had grown would once more support the flame of a candle. He was within reach of the fundamental discovery of biological chemistry, the carbon cycle, but he went no further.

While trying to make carbon dioxide by heating limestone in a gun barrel, he got a large quantity of a gas which burned with a blue flame. This was

carbon monoxide, but Priestley brushed it aside with
the remark that it must come from the remains of
the animals whose skeletons formed the limestone.
The credit for discovering carbon monoxide goes to
Cruikshank.

So it was with all Priestley's work. He was a wonderfully alert observer. He was accurate and pains-
taking. He was ingenious to the point of inspiration.
But almost never did he draw a correct conclusion.
The phlogiston theory always got in the way. Priestley's mind was free—free from dogma social and
religious. But it was not free from the curious human
craving for intellectual bonds. He was like an old
circus horse which, when turned out to graze in a
wide pasture, imagines that the ring still encloses
him and gallops sedately around a small and perfect
circle at the center of the field. Priestley rejected
the Athanasian Creed and all other creeds. He rejected the political philosophy of the age of George
the Third. But when he attacked chemistry, his real
life work, he felt the need of support from the past,
and he chose the phlogiston theory, which was even
then showing unmistakable signs of decay.

Science needs the Priestleys. It needs the skillful
fact-fishermen. But it also needs the tough-minded
thinkers like Boyle and Lavoisier who can discard
the obsolete theories of the past and build up new

structures of thought on new and firm foundations. Priestley never did this. He was satisfied with the past. He thought the phlogiston theory needed only a little strengthening to stand forever. And when it did fall, Priestley almost alone among men of science remained faithful. The arch heretic, denounced for his radicalism from every pulpit in England, clung with pathetic loyalty to a doctrine which was much more completely dead than any of the religious creeds he'd tried so hard to destroy.

These are hard words to apply to so great a man. "Conservative" and "orthodox" are the most insulting epithets in the vocabulary of science, for science is like a stalk of coral, alive only at the growing tip. And it's the height of irony that these expressions should come to be applied to Priestley of all men. Only in the matter of phlogiston was he dogmatic. In other fields, religious and political, he was as radical as any man in England. He was attacked for his opinions by pulpit and press. And he was finally driven into exile to end his career three thousand miles from the center of the "philosophical life" he loved so much. To be at peace with Priestley's kindly and earnest ghost, we'll have to consider the other side of his curious dual personality, the radical side of him which almost cost him his life.

Priestley, like Boyle, lived during a period of

crisis. The French Revolution was just over the horizon, and it cast its dark shadow a long way ahead. The industrial revolution was just beginning. Religion was being challenged fundamentally—as it had never been challenged before. And the structure of government, which had remained essentially unchanged for nearly a century, was being savagely attacked on numerous fronts. All along the line the upholders of the established order were nervous— the bishops, the landowners, the great merchants, the members of Parliament, and the king himself. And well they might be, for presently they'd look across the channel and see their fellows in France strung up on lamp posts, torn to pieces in the streets, or driven into exile.

On July 14, 1789, came the destruction of the Bastille, and a shudder of apprehension ran through the governing classes of Europe. Guilty consciences awoke and resolved to be more guilty in future or they'd never get a chance to be guilty again. Garrisons were strengthened in populous centers. New laws against sedition were enacted and the old ones polished up. The press was watched more closely. And very useful was that tried-and-true weapon of reaction, the appeal to religious prejudice.

In 1780 Priestley had left Lord Shelbourne, who couldn't quite stand for some of his ideas, and moved to Birmingham to preach to the congregation of the New Meetinghouse, one of the most liberal in England. By this time he had earned the well-deserved hatred of the Established Church. The Church did not fear the Deists and Agnostics who denied religion altogether, for it knew very well that these doctrines had no popular appeal. But Priestley it did fear and with good reason. His Unitarian doctrines were just what it didn't want to see established, for Unitarianism, while appealing to the fundamental human craving for religion, denies the existence of those mystical details like the Trinity and the Sacraments which are the chief support of priestcraft.

If Priestley had stuck to science and theology, he would probably have escaped physical violence at least, for the government after losing the American war was in no mood to crusade in favor of the Church. But in those days theology led directly into politics by way of the Corporation and Test laws which excluded all non-members of the Church of England from voting or holding office. Priestley led the attack on these laws with the characteristic vehemence which he showed when a matter of principle was involved, and it wasn't long before he was also denouncing the rotten system of representation in Parliament which kept such laws on the books.

These activities brought Priestley much notoriety

and disapproval from on high. But for a time all was serene. He preached a little, experimented a little, and wrote volumes on science, religion, and politics. He was free from financial worries, he was courted by the great and the less great, and he was recognized as a prominent leader of liberal thought.

But after the fall of the Bastille, all this was changed. When there's revolution in the air, the mildest dissatisfaction with conditions as they are begins to look like sedition. The Dissenters, the Catholics, the embryonic labor unions, and the philosophical friends of liberty in general all came under suspicion. And Priestley naturally among them. He began to get threatening letters. Some of his weak-kneed parishioners stayed away from church when he preached. And when he traveled down to London he noticed a certain coldness among the members of the Royal Society.

In Birmingham itself the feeling of tension increased rapidly until July 14, 1791, when the crisis arrived. For some days the city had been in an uproar over an inflammatory handbill which had been found in a public house and circulated widely by Priestley's enemies. Priestley, of course, denied any knowledge of it, but the harm was done. The lowest classes of the population were convinced that the Dissenters and Liberals were planning a massacre,

or at least the destruction of organized government
and religion.

For July 14th the Liberals of the city had planned
a dinner in celebration of the second anniversary of
the Fall of the Bastille. It was to be a great feast of
oratory in honor of British liberty as compared to
the conditions which had brought on the French
Revolution, and many respectable and conservative
citizens had planned to attend. But as the date drew
near, the list grew smaller and smaller as the more
timid made their excuses, and after the incident of
the handbill the promoters decided to call it off.
They prepared a notice to this effect, but changed
their minds at the very last minute on receiving
assurances from the owner of the hotel where the
dinner was to be held that no trouble was likely.
They didn't suspect until later, nor was it ever
proved, that the hotel keeper had made his peace
with the instigators of the riot. But the rioters did
not attack the hotel. They waited in very unmob-
like silence until the banquet was over, and didn't
start their work until a leader raised the rallying cry
of "Church and King."

★        ★        ★

Priestley did not attend the dinner. He had been
warned by his friends that his presence might make

trouble. That evening he sat at home playing back-gammon with his wife. It was a mild occupation which took perhaps one tenth of his attention and one hundredth of his intelligence. He threw the dice with an appearance of interest and made the simple moves authorized by their faces, but his mind was far away from Birmingham, from his wife, from backgammon. Just then he was thinking of a scene in Cracow two hundred years before when Faustus Socinus, an early Unitarian, had fled from the city before a mob and looked back to see the long yellow flame from his burning house proclaiming to the cornfields of Poland that Christ after all *was* a God.

Birmingham was a long way from Cracow, and two hundred years was a long time, but, thought Priestley, he'd been hearing that word "Socinian" rather often of late. He was used to being called names from the pulpits, and the things the clergymen said about him had long since ceased to matter. But "Socinian" from a crowd on a street corner had a different sound. Ominous, like all words which the mob snarls out but does not understand.

"Someone," thought Priestley, "must be teaching the people that 'Socinian' has a hateful meaning, that the poor dead word is alive and powerful, fit to mingle with catcalls and hisses; that a man who wears it is a traitor, a blasphemer, and a rebel."

He threw his dice again and moved his counters on the painted board. From the back of his mind came a Latin verse which formed the epitaph on the tomb of Socinus:

*"Tota licet Babylon destruxit tecta Lutherus Muros Calvinus, sed fundamenta Socinus."*

("Luther destroyed the houses of Babylon, Calvin the walls, but Socinus uprooted the foundations.)

"You'd think," thought Priestley, "that these Protestant English would applaud a man who destroyed the mystical foundations of the Roman edifice they hate so intensely. But no. That's not the way things work. They want the foundations left intact, so that they themselves may build a comfortable structure on them—and be comfortable deans and bishops within it."

Priestley shook his head sadly and threw the dice again. "The way of the transgressor may be hard," he thought, "but it's nothing to the way of the man who thinks for himself. After all, he's the real transgressor. He breaks great rules while the thief and the adulterer break little ones."

Priestley threw the dice again, but before he could move his men a frantic peal came from the bell in the hall. Doorbells have their language. They can express urgency if nothing else. Mrs. Priestley looked up with a start, and Priestley went to the door without waiting for the maid to open it.

The visitor who'd nearly pulled the bell handle from its socket was Samuel Ryland, a well-known Liberal and one of Priestley's staunchest supporters. He wore no hat and was out of breath.

"Hurry," he panted, "they've burned the New Meetinghouse. They're burning the Old Meetinghouse. Then they're coming here."

" 'They'?" asked Priestley. "Who are 'they'?"

"The mob. The people. They waited outside the Memorial Banquet for you to come out, and now they're running around the streets with torches and clubs."

"But I have no quarrel with the people," said Priestley. "I've spent the better part of my life fighting their battles here and in London."

"They're crying 'Church and King.' They're coming to burn your house. And if they catch you in it, they'll burn you too, friend or no friend."

Mrs. Priestley seems to have had more presence of mind than her husband, or perhaps less faith in human justice. While he was standing dazed and wondering in the doorway, she gathered up what money there was in the house, found his coat, threw it over his shoulders, and pushed him into the street.

All that night the cry of "Church and King" was heard in the the streets of Birmingham. Priestley's house was burned and the walls pulled down with wrecking hooks. His laboratory was destroyed, too, with the precious apparatus which could not be duplicated in all Europe. The houses of his friends went next, even their country estates many miles from the city. Priestley himself fled to London where he was safe if not welcome. The authorities tolerated no mob, however friendly, in the capital.

The mob held Birmingham unopposed from Thursday until Sunday. Priestley's friends tried to find the magistrates, but they had disappeared. The city police were gone too. There were some feeble attempts to organize a volunteer constabulary, but they had little support and came to nothing. Late Sunday evening a troop of cavalry rode calmly into the city. The mob dispersed without resistance. A pleasant time had been had by all.

Of course the authorities disclaimed all responsibility for the riot, and to give weight to their words they hanged a few of the more obscure rioters. But the facts were against them. The mob was not the usual senseless, howling thing which emerges from the slums of a great city. It was organized. It knew just what to do. It carried hooks and ropes for pulling down walls, and straw and torches for firing buildings. It was wholly unarmed. A whiff of gunpowder would have scattered it. Within a day's hard ride from Birmingham were plenty of troops, and the constables of those days went armed to the teeth. But not a shot was fired. The mob with its hooks and torches held Birmingham for three days and three nights until every house and chapel which had been marked for destruction was reduced to ashes and rubbish. Only then did the troops appear.

Priestley got a great deal of sympathy from his friends and from Liberals in general, but little good did it do him. The Birmingham riots had identified him in the popular mind with the enemies of Church and State. Wherever he went he was followed by threats and abuse. Handbills were scattered about the streets of London describing him as a rebel and a traitor and demanding his death. He was burned in effigy. His servants were afraid to stay with him. His neighbors moved away in fear of another riot. His mail every morning brought letters comparing him to Guy Fawkes, to Danton, to the devil, and threatening to boil him in oil. His fellow members of the Royal Society avoided him like the plague, and in the pulpits of the Established Church his wickedness and blasphemy were favorite subjects for impassioned sermons. The Rev. Dr. Tatham, rector of Lincoln College, Oxford, was very emphatic:

> "Long have you been the Danger of this country, the Bane of its Polity, and the Cankerworm of its Happiness. Long, too long, have your Principles tended to bereave it of its Religion, its Constitution, and consequently its King."

Even Burke, the great champion of American liberty, turned against him and denounced him in the House of Commons. England was no place for Priestley. He decided to emigrate.

On April 8, 1794, Priestley and his wife took ship from London. He went with numerous regrets, for he was leaving the only life he knew, the life of philosophical contemplation in the midst of a civilized community, and he was going to a country which had not yet passed entirely beyond the pioneer stage. But America of the late Eighteenth Century was the only refuge for such men as Priestley. Washington was still President, and the principles of the Constitution and the Declaration of Independence were still in force. A man like Priestley could be sure of a hearty welcome and a tolerant attitude toward his not-very-violent ideas.

He landed in New York on June 4th and was welcomed by an enthusiastic crowd at the dock. There were addresses by prominent men and banquets in his honor. When he reached Philadelphia he was honored by the American Philosophical Society and offered by unanimous vote of the trustees the professorship of chemistry at the university.

But Priestley was sixty-one now and discouraged. He preferred to retire with his sons to Northumberland on the Susquehanna where they were founding a colony of English exiles. He went to Philadelphia several times to lecture and preach, but always returned to Northumberland where he found the peace his failing health required. He wrote much on theology and did a little research, but his life work was over.

Only once did he regain any of his old scientific vigor—when he made a last desperate attempt to revive the beloved phlogiston theory with a pamphlet called "The Doctrine of Phlogiston Established." Needless to say, it came to nothing. Phlogiston was too dead to resurrect. But Priestley never lost faith in his one dogma. He died in 1804 at Northumberland—the prophet of a doctrine honored by all but its discoverer, a father of the chemical revolution who did not recognize his child.

# Henry Cavendish, the Measuring Machine

# CHAPTER VIII
## Henry Cavendish, the Measuring Machine

THE Royal Society Club had just finished dinner, and the members were standing about with the air of great men who thoroughly appreciate their own greatness. One member, it was clear, was not enjoying himself. He looked nervous, glancing from face to face, and noting with alarm that many were strange. He wore very curious clothes, rather like a family portrait by an indifferent artist, and he shuffled his feet like a small boy who wishes he were somewhere else. The other members appeared to be paying very little attention to him, but when he did speak, which was rarely, a great silence fell until he was done, which was soon.

The room was crowded. If anyone wished to get from one end to the other, he had to move in a viscous medium of men so interested in what they were saying that they didn't want to stand aside. From a far corner a little convoy was approaching —a large pompous gentleman followed by another gentleman rather more stiff than pompous. It pushed through the crowd, setting up a disturbance ahead of it like a ship with a very blunt bow. The nervous man in the funny clothes detected its approach from afar and looked even more nervous than before.

At last the convoy reached its destination. The pompous man, who was Dr. Ingenhousz, addressed the nervous one, who was Henry Cavendish.

"Mr. Cavendish," he said with ponderous formality, leading forward his companion, "this is the Baron von Plattnitz, a man as well known in his native country, which is Austria, for his philosophical attainments as for his noble birth. He has asked me for an introduction to you, sir, a pleasure which he has long anticipated. Allow me to present the Baron von Plattnitz."

Cavendish looked down at his feet, fumbled the large buttons on his coat, and made curious motions with his head and shoulders. He said not a word. The Austrian gentleman came forward and bowed profoundly.

"Mr. Cavendish," he began, "I came to London not for business. I have no business in England. Nor for pleasure. My own Vienna is much more gay and amusing. I came solely for the privilege of conversing with you, whom I consider one of the brightest ornaments of our age and one of the most illustrious philosophers the world has ever produced. Now that I have had the pleasure of meeting you——"

He stopped suddenly with open mouth. Cavendish was gone. He had spied an opening in the crowd and popped through it like a rabbit through a hole in a hedge. He dashed out the door and down the steps. He jumped into his carriage and was driving like mad for his home at the other end of London.

★   ★   ★

The Honorable Henry Cavendish was a great aristocrat, descended from numerous Norman conquerors, connected with royalty in various ways. Dukes and earls roosted thick as blackbirds in his family tree, and he possessed that crowning glory of aristocracy—tremendous wealth. But never was a man less like the traditional great lord. Lacking were the fine clothes, the cultivated tastes. Lacking were the polished manners, the inbred arrogance. Lacking were all social interests, all concern with public affairs. He lived a life which would have been considered uneventful by the dullest night-watchman. He fortified himself against social contacts. He hated praise and resented compliments. He became one of the most famous men of his time, but he got no thrill out of it. What he did in science was for his personal amusement alone, just as other men in his position

killed grouse and broke hearts. The advancement of
science mattered not at all to him. He just didn't
care. He was a measuring machine.

Cavendish was born in 1731 at Nice, where his
mother had gone for her health. It might have been
a different date or a different place, for neither date
nor place had any effect on Cavendish. If he'd been
born a hundred years earlier in Nova Zembla, we
can be sure that this coldest of warm-blooded
creatures would have lived just the same, measuring
the ice and desolation of Nova Zembla in the same
spirit as he measured the physical and chemical
qualities of England.

He went to school at Hackney. But we know no
more than the plain fact. He left no memory or
tradition behind him. He never confided to anyone
the emotions and struggles of his schooldays. There
probably were none. He studied at Cambridge for
four years and left without a degree, perhaps because
of the religious test required before a degree could
be conferred. Cavendish had no religious opinions,
and the pale glitter of an M. A. after his name was
not sufficient bait to draw from him even the feeblest
lie. He left Cambridge in 1753, went to London, and
disappeared from all public notice until the fame of
his scientific discoveries dragged him into unwilling
notoriety.

There have been many attempts to work out by hook or by crook what manner of life Cavendish led before his fame made him a public character, but all are alike in their total failure. We don't even know whether he lived in affluence or in poverty. Both were the same to him. When, through the death of relatives, he became one of the richest men in England, his habits changed not a bit. He still lived and acted as if he had just enough for the simple minima of life.

Almost the only relics of this period of Cavendish's life are the laboratory notes which he accumulated in tremendous quantities. They prove that from the time he left Cambridge he was hard at work. Measuring, always measuring. He investigated the chemical reactions of arsenic and weighed carefully all the ingredients he used. He determined the freezing point of mercury, and developed a thermometer which would measure this better than any had done before. He carried on other researches on heat, recording his observations but seldom putting them together into anything resembling a theory. His first appearance before the public was when, in 1766, the Royal Society got hold of his paper on "Factitious Airs" and published it in the *Philosophical Transactions*. From then on he was an unwilling celebrity.

This paper contained the first accurate investigation of hydrogen, the element which was to give Cavendish his chief importance as a chemist. And it showed what his mind was like. There is no theorizing, no putting together of observations. Only measurement. Cavendish went about getting his hydrogen with painful exactness. He found what metals would yield hydrogen when treated with acids and he found exactly how much each would yield. An ounce of zinc gave 356 measures of the gas, an ounce of iron 412, and an ounce of tin 202. These observations might have grown into the "theory of combining weights," but Cavendish's mind did not work that way. He verified his figures and passed on to determine with equal care the specific gravity of the hydrogen he got. His results were not very accurate, because the practical difficulties of weighing this extremely light gas were great. But they were the best yet made. Having measured his hydrogen, Cavendish then measured its inflammability by exploding it with different volumes of air. His results are expressed in definite figures, but he did not notice the moisture which resulted. He was looking for one thing only, the amount of hydrogen which had to be mixed with the air to make it burn. He found this, but it wasn't for nearly twenty years that he discovered what product was formed.

In this paper, and all his other papers, Cavendish

*H. Cavendish*

THE HONOURABLE HENRY CAVENDISH,

### HENRY CAVENDISH
Eldest son of Lord Charles Cavendish, fabulously wealthy, a hater of women, and possibly the most important chemist of his time.

used the language of the phlogiston theory. Its terminology was widely understood and convenient. But
he could hardly be called a partisan of the theory.
Priestley accepted the phlogiston theory and fought
for it with a loyalty seldom lavished on anything but
a religious doctrine. But Cavendish was not interested in controversy. He made quantities of hydrogen,
measured it, investigated its properties, and let others
worry about the implications of his discoveries. If
you'd asked him which side he took in the phlogiston
controversy—the burning scientific question of the
day—you'd have got a blank look or no look at all.
He used the phlogistian terminology because it was
convenient, but that was as far as he went. The gas
would weigh the same and act the same whatever he
called it.

It is hopeless to attempt to give a chronological
account of Cavendish's discoveries. He didn't care
what the world thought of his work, and so he took
no pains to get his results in print. Many of his
papers were not published until long after his death,
and many were not written down at all, remaining
a mere mass of figures in his notebooks. Often a discovery of importance would lie untouched in his files
while other men worked hopefully on the same
problem. So few were his contacts with the world
outside his laboratory that he seldom paid the slight-

est attention to what his fellow scientists were doing. Even when he learned by accident that they were working on a problem he had already solved, he seldom took the trouble to tell them.

The account of such an isolated life is bound to be a series of fragments. Cavendish was busy all the time, but he apparently worked without any end in view, without any intention of making his work valuable to science. He jumped from chemistry to physics, from mathematics to astronomy, from the electrical apparatus of the torpedo fish to the ancient calendar of the Hindus. When a problem ceased to interest him, he'd drop it. What did he care if the world was waiting for the solution? That was none of his business.

A few acquaintances did manage to penetrate to his laboratory and watch him at work. They all made the same report, that his success was due to an extraordinary accuracy and a marvelous capacity for taking pains. His apparatus was clumsy-looking, without finish or elegance, but its essential part, the balance or thermometer, had been fussed with until it was as accurate as human hands could make it. Every experiment was performed fifty or a hundred times before the results were averaged. Every bit of material was tested before it was used. Every disturbing factor was allowed for with painful care.

At length a figure would emerge, a figure of many decimals. Cavendish's work was over. He turned to the next experiment. Perhaps he had passed within an inch of some important law of Nature. He didn't know or care. Others could generalize. He measured.

The greatest achievement of Cavendish was, of course, the discovery of the compound nature of water, which has been discussed in a previous chapter. But this was only a small part of his work. He did a great deal with heat, especially its effect on liquids, and in the course of this research developed a number of vastly improved thermometers. He analyzed for the first time the hard waters of the London wells. He developed the first eudiometer for measuring the percentage of oxygen in air. His notes bristle with observations which he jotted down in passing. In trying to freeze mercury, for instance, he found that a freezing mixture of nitric acid and snow would give a lower temperature if the acid were slightly diluted. He took account of this in making his mixtures, but it didn't occur to him to investigate the definite hydrates which we now know nitric acid and other substances form with water.

Perhaps the most interesting experiment Cavendish ever did was his famous feat of weighing the earth. Here again is illustrated the essentially quantitative and non-theoretic nature of his mind. The attempt to weigh the earth was nothing new. Ever since Newton demonstrated the gravitational relations between the members of the solar system, astronomers had yearned for an accurate determination of the earth's mass, for on this figure depended most of their calculations. Several methods had been tried or proposed. There were the "mountain method," the "tidal method," and finally the "torsion balance method," which was the one Cavendish decided to use.

By the time he took hold of the problem it had been discussed at length, and all the theoretical thinking-out had been done. Even the apparatus to solve the problem had been designed by an ingenious clergyman named Mitchell who had died before he could try out his idea. Nothing remained but the measuring, which was just what Cavendish loved. He attacked the job with a cold emotion which in another man might have been called enthusiasm.

Mitchell's apparatus for weighing the earth is very simple in principle. It depends on the fact that the earth is not the only body which exerts gravitational force. Every body attracts every other body, the strength of the attraction depending only on their mass and the distance between them. Thus if the attraction between two small bodies at the surface of the earth can be measured, the mass of the earth can be derived from this figure. We know the force with which the earth attracts a body at its surface. This is the *weight* of the body. We know the distance between the earth's surface and its center. The only remaining unknown factor will be the mass of the earth, and this can be found by solving a simple equation.

The apparatus which Cavendish used for measuring this force consisted of a wooden beam suspended at its center by a fine wire and carrying a small lead ball on either end. Two large lead balls were placed in such a way that in attracting the small ones they tended to turn the beam against the resistance of the wire's elasticity. The angle through which the beam was turned could be measured, and the force necessary to twist the wire could be computed mathematically by swinging the beam and timing its vibrations. The stiffer the wire, the faster it would swing. The rest was arithmetic.

In theory the problem, once grasped, is very simple, but the technical difficulties are tremendous. The force to be measured is so small that it looks like a decimal point followed by a row of zeros as long as a bead necklace. The slightest disturbing influence such as a change of temperature throws the whole thing out of gear. But this was ideal recreation for Cavendish. He worked at the problem for years, performed the experiment over and over again, and finally emerged with the conclusion that the density of the earth was 5.48. This result was so accurate that not until forty years later could the combined resources of the Royal Astronomical Society find a better figure.

\*     \*     \*

How did Cavendish live while he was performing these mighty but unimaginative *tours de force?* Who were his friends? What were his amusements? What were his avocations? The answer is simple. He had none. We have called him a measuring machine, and a measuring machine he was. His life was about as eventful as that of one of his thermometers. He had no opinions on general subjects; he expressed no emotions—unless an all-inclusive dislike of humanity may be called an emotion. If he had some secret defect or deformity which made him avoid his fellow men, he carried that secret successfully to the grave, leaving no material for the modern psychological grave robber to work with.

His favorite residence was a large estate in Clapham, then an outlying suburb of London. The whole house and grounds were filled with a miscellaneous collection of instruments ranging from delicate thermometers to a full-sized blacksmith's forge set up in one of the living rooms. The top floor was an observatory containing a large telescope, and a high platform built in a clump of stately trees served a similar purpose. Along the halls stood racks of glass vessels and reagent bottles. A visitor had to step carefully to avoid knocking over an elaborate microscope or a carboy of sulphuric acid.

To the educated inhabitants of Clapham Cavendish was an eccentric celebrity; to the ignorant he was a wizard and his house a den of black magic. Strange smells would drift through the open windows, and dull explosions would waken the neighbors from sleep. The servants were silent, uncommunicative creatures who seemed to have absorbed some of the character of their master. The inquisitive couldn't learn much from them about the daily life of Cavendish.

But so well known a celebrity can't keep entirely out of sight. His very defenses are a challenge to the curious. Quite a body of anecdote grew up around this coldest of human beings.

A man with such a passion for accurate measurement would be apt to arrange his life according to a schedule. Cavendish did. His daily routine was as accurate and unvarying as the graduations on one of his thermometers. He rose by the clock, took his meals by the clock, and retired by the clock. Even his single recreation, driving in his carriage, was fitted accurately into this schedule. To the rear wheel he attached a clumsy wooden instrument called a way-wiser which counted the revolutions of the wheel and so measured the distance covered. Cavendish knew each day exactly how far he intended to drive. He kept his eyes on the way-wiser. When it registered half the predetermined distance, he called to the driver, who turned around and went home by exactly the same route. The day's recreation was over.

This passion for regularity extended even to his clothes. He never owned more than one suit at a time. The pattern and cloth never varied, and he knew exactly how long a suit might be expected to last. When the date for discarding it approached, he sent for his tailor and ordered another exactly like it. At the end of his life he was wearing clothes which, although new. were at least forty years out of style.

His social contacts were almost zero, but he did go to meetings of the Royal Society Club and to an occasional dinner at the house of some learned acquaintance. Sometimes at the sight of an unknown face he would bolt for the door and be gone before his host knew what had happened. On even more rare occasions he would give dinners himself to a chosen two or three. His style of entertainment was not lavish. The food never varied—a leg of mutton, nothing more. Once he invited the unprecedented number of four. His butler asked what he should serve.

"A leg of mutton, of course," said Cavendish.

"But that won't be enough for five."

"Two legs then."

His dislike for his fellow men was nothing compared with his loathing for women. There were female servants in his house, but they had strict orders to remain where he would not see them. If he as much as caught sight of one, she lost her job. When he passed a woman on the street, he looked fixedly the other way as if she were a Medusa intent on turning him to stone. When he went to walk, he did so at night because he had learned that two young women of the neighborhood had timed his arrival and used to watch him getting over a stile.

After one of the dinners of the Royal Society Club the members were standing around the room discussing the things which philosophers are supposed to discuss. One of the members happened to go to a window and notice that a very pretty girl was looking down from the upper story of the house opposite and observing the philosophers with a friendly smile. At a signal from him the other members crowded to the window to admire the fair creature who was so obviously admiring them. Cavendish saw the gathering and, thinking that his learned colleagues must be looking at the moon or something else equally worthy, shuffled over to the window. He took one look, saw the smiling girl, muttered a disgusted "Pshaw," and hastily retired to the back of the room.

Such a man would not be likely to sit for a portrait, and Cavendish never did so knowingly. But a portrait of him exists nevertheless. A well-known artist of the time asked Sir Joseph Banks to persuade Cavendish to sit for him. Sir Joseph laughed heartily. It had been tried many times before without success. But the artist persisted, and finally Banks agreed to invite him to a dinner at which Cavendish would be present and seat him where he could sketch him unobserved. The view wasn't perfect. For a while all the artist could do was to draw the gray-green coat and the antique three-cornered hat. At last Cavendish turned his head. The artist's pencil worked like mad. Cavendish turned back again, and the artist fled to safety before his sketch could be confiscated.

When Cavendish was in early middle age the deaths of various relatives made him one of the richest men in England. This was a great trial to him. He bought all the apparatus he could use, but this considerable expense didn't make the slightest dent in his income. Year by year his fortune increased. His bankers managed the money, collected the interest and rents, and knew better than to bother him about them. They invested some of the accumulated funds in standard securities, but finally they could stand it no longer. They had a conference and decided to send their bravest employee to beard the lion in his den. The unfortunate man who was given the dangerous task knocked at Cavendish's door and asked to see the master of the house.

"You can't see him," said the butler. "He's in his laboratory."

"But I must see him. It is very important."

"You can't see him. I have orders not to admit anyone."

"I must see him," said the banker, and pushed past the door.

He found Cavendish in his laboratory wearing a tremendous frown.

"Who are you, and what do you want?"

"I was sent by your bankers, and I want to report that you have an uninvested balance of 65,000 pounds. That is a large sum to leave idle, Mr. Cavendish. What shall we do with it?"

"Don't bother me again, or I'll take the money

away from you and place it with someone who won't annoy me."

"Shall we invest it, Mr. Cavendish?"

"Yes, yes. But don't come here again." He slammed the laboratory door in the banker's face.

★     ★     ★

Bacon, Paracelsus, Boyle, and Priestley were all intensely religious, each in his own way, but religion bothered Cavendish not at all. He couldn't measure it; he couldn't multiply the Trinity by itself and get a perfect square. So religion played as small a part in his life as poetry, love, or any other imponderable. No one ever heard him refer to God, to Christ, or to the Bible. He wasn't even anti-religious. He just didn't care.

When a man is about to die, when he realizes that he himself is about to become imponderable if anything, his religion is apt to burst into belated bloom. But not so with Cavendish. At the age of seventy-nine he decided that the end of his life had come. He was weak and worn out. The life force had been steadily failing. Cavendish plotted its descending curve and computed just when it would touch zero. He called for his valet.

"Mind what I say. I am going to die. When I am

dead, but not until then, go to Lord George Cavendish and tell him of the event. Go."

Half an hour later the bell rang again. The valet answered, and in a weaker voice Cavendish ordered him to repeat his instructions.

"When you are dead," repeated the valet, "but not until then, I am to go to Lord George Cavendish and tell him of the event."

"Right," said Cavendish. "Now go."

The valet retired, and Cavendish turned his face to the wall. A little later the valet tip-toed back. His master was dead as he had predicted.

★    ★    ★

So lived and died the coldest, most unhuman mortal who ever wrote his name large in the history of science. Small men are sometimes cold, but great ones very seldom. They need human aspirations and desires to drive them toward achievement. But Cavendish was driven by no such spur. He had more wealth than he could use. He had birth and position. The respect of his colleagues was the only thing which he would have had to work for, and this he wanted less than anything else in the world. What he did in science was done for his own amusement alone. He was a measuring machine. His sole interest was to measure the objects in the material universe around him. He contributed greatly to the advancement of every science known at the time, but he would have been just as well pleased if his discoveries had died with him and his precious notebooks had followed him to the grave.

# Antoine Laurent Lavoisier, the Grand
Seigneur of Science

# CHAPTER IX

## Antoine Laurent Lavoisier, the Grand Seigneur of Science

SOME children are born with a silver spoon in their mouths, some with a gold one. Lavoisier must have been born with a platinum crucible in his, for he started rich, became richer, and also became the most famous chemist of his age—an achievement almost unique in the history of science, for scientific ability and money-making ability seldom exist together in the same individual. The practical calculations of commercialism are very different from the abstract calculations of science. And besides that, the scientist seldom wants money badly enough to pay for it the necessary price in time and effort. The things he desires money won't buy him.

But Lavoisier lived in France under the Old Régime, and the social-economic set-up was not at all like what it is to-day. Almost every opportunity to make money was directly or indirectly in the hands of the government. There were monopolies and concessions. There were pensions and annuities.

There were direct grants from the crown. There were a thousand different kinds of graft. And naturally, since the government was controlled by a small governing class, these various plums fell into the laps of the members of this class. Lavoisier was an aristocrat. He was expected to become rich—and did.

The system of special privilege under the Old Régime had many and obvious disadvantages. It supported in unproductive idleness a vast number of parasites, and it imposed a discouraging burden on the lower classes, but it had one advantage which almost made up for its faults. It freed men like Lavoisier from the necessity of developing the selfish psychology of commercialism and allowed them to work for the public welfare if they were so inclined. Lavoisier was the shining example of what special privilege can do for society. He made up for a dozen silly courtiers strutting like satin-clad peacocks up and down the halls of Versailles.

Lavoisier was born in 1743 at Paris. His family held no title and owned no lands, but it had risen step by step during the last hundred years from the humble rank of postilion in the king's service to the top of the legal profession at the capital. Lavoisier's father was Sheriff of the Court of Justice—a position of dignity and importance. He married the heiress, Emilie Punctis, and was in a position to see that his son got a good start in the world.

In those days sons were apt to follow in the footsteps of their fathers, and for several generations the family profession had been the law. But the elder Lavoisier soon saw where his son's talents lay and sent him to the College Mazarin, famous for its scientific faculty. Here he got the best scientific training available, which wasn't so very good, and was ready at the age of nineteen to please his father by attending the law school. He took his degree in a year and got his license to practice a year later, but the inoculation didn't take. No doubt he would have been a good lawyer, but there were other and vastly more fascinating things to do in Paris in 1763.

Among the other things there was literature. Lavoisier was no dry-minded technical man. His interests covered the whole spectrum of life. His first youthful effort was a play which was to be called *La Nouvelle Héloïse*. Perhaps he didn't know the title was not exactly original, or perhaps he was trying to make a play out of Rousseau's famous novel. The world never learned, for when he'd finished only the first three acts he became fascinated by the chemical lectures by Rouelle at the Jardin du Roi which had captured the imagination of fashionable Paris.

Rouelle does not rank now as a great chemist. His teachings contained much that was wrong, and his independent discoveries were negligible. But he must have been a wonderful lecturer, for in silks and powdered wigs the great people of Paris crowded to hear him. He dressed like a dancing master in velvet coat and huge wig, but as soon as he got into the swing of his lecture, the niceties of dress and manner went by the board. Off would come the coat to lie forgotten in a corner. The wig would follow, and little Rouelle would stand in his close-cropped gray hair amid clouds of evil-smelling smoke demonstrating to the élite of the capital the mysteries of combustion and calcination. These lectures became the fad of the day. All Paris was thrilled by what it thought was a new insight into the mysterious ways of Nature. Lavoisier's literary ambitions died a sudden death, and he began the first of his long series of scientific researches.

Lavoisier's first enterprise in science shows determination but not imagination. Perhaps as a new convert from literature he thought he should learn scientific discipline. At any rate he set up a number of barometers in his grandfather's house and recorded their readings several times a day. His aunt and his cousin assisted, and his father bought the instruments and arranged for correspondents in various provincial cities. The readings were continued for thirty years, but no use was ever made of the data. By that time Lavoisier had other things to worry about.

From this abortive attempt at meteorology Lavoisier passed on to geology and began studying exhaustively the rocks and strata in the neighborhood of Paris—especially the gypsum whose abundance near the city gave plaster-of-Paris its name. This was what led him into chemistry, for after he'd mapped the quarries from which the gypsum was taken he proceeded to find out what made the burnt gypsum set with water into the hard plaster so useful in building and the arts.

The experiments with gypsum were successful and important, but what makes them especially interesting was that Lavoisier in this work was almost the first chemist to use an entirely *quantitative* method. He heated the gypsum, found that the vapor given off was pure water and that it was exactly equal in weight to the water absorbed by the plaster when it hardened. He decided that the setting of the plaster was due to a recrystallization and that it was the water combining into new crystals with the powdered and burnt gypsum which hardened it into plaster-of-Paris. In his very first chemical research Lavoisier realized the importance of weight, and it was this practice, which he continued throughout his career, that was responsible for his greatest discoveries.

In 1765 the Academy of Sciences offered a prize for the best system of street lighting, and Lavoisier jumped at the chance with enthusiasm. One of his fundamental convictions was that the real function of science was public service, and he was anxious to prove that as a scientist he could contribute something valuable to the public welfare. He went about it with characteristic, even fanatical thoroughness, shutting himself for six weeks in a dark room so that his eyes would become sensitive enough to distinguish between the effects of various lamps and reflectors. The report he submitted was a marvel of completeness. He considered every known type of lamp and candle, every known type of reflector and lamp post. Since the prize was to be awarded for elegance and economy as well as for efficiency, Lavoisier drew dozens of neat designs. His decision that candles were on the whole more effective and economical than lamps seems a bit strange to us, but at any rate he won the prize and received a gold medal from the king. He was only twenty-two, and already he'd performed the first experiment in quantitative chemistry and had done the first of a long list of public services.

This triumph gained him so much renown that in

1768 he was proposed for membership in the Academy and elected the following year, the youngest member of that august if somewhat pedantic body. He at once plunged into the work of the association with an energy and devotion almost miraculous. During the twenty-five years of his membership he drew up more than two hundred reports on subjects ranging from the water supply of the city to mesmerism and the divining rod. He could finish an intricate report while his colleagues were going through the preliminary dinners and discussions.

About this time Lavoisier decided that he'd have to have more money. Science, as he saw it, wasn't a mere laboratory job—one man surrounded by apparatus and notepaper prying loose the unwilling secrets of Nature. It was a kingdom and he proposed to be king. He'd need assistants and secretaries. He'd need correspondents and field workers. He'd need artisans to make unheard-of instruments. And he'd need the friendship of every other scientist in France. All this would take money, a great deal of money. Well, why not? The government controlled most of the money in France. It made a practice of favoring men of culture and accomplishment. And it didn't look too closely into the details of finance. Lavoisier decided to join the Ferme Générale.

This was a step which had a profound influence on his later life, for the Ferme Générale was the best-hated branch of a most unpopular government. In its long career of more than four hundred years it had collected such an accumulation of popular dislike that by the time of Lavoisier the mere mention of its name in a public house drew a volley of threats and curses.

Back in the beginning of the Fourteenth Century the current king had decided that the finances of France were rather a bore, not worthy of his time and attention. So he'd sold the revenues of the kingdom to a group of bankers and let them worry about collecting. Whatever they could get above the sum they paid, they could keep, and if he heard they were keeping too much, he'd charge a higher price the next year.

From such naïve beginnings developed the vast system of the Ferme Générale. As France grew, it grew also. When expensive wars occurred, and they were always occurring, new taxes were levied and sold in advance for a lump sum. Monopolies were granted, such as salt and tobacco, and sold each year. Customs barriers were erected between the provinces, between each city and the surrounding country, and every turnip or cabbage had to pay tribute to the Ferme Générale.

Each year the Ferme had to pay a higher price for its privileges, and each year it had to squeeze the people a little harder if it hoped to clear expenses. By 1768, the year Lavoisier entered the corporation, the country was nearly desperate. Whole regions were in chronic revolt. Smuggling was on a vast scale. The Ferme maintained a standing army of its own to protect its tax collectors from violence. When the common people passed its headquarters in Paris, they spat and swore. All the hatred of France for its rotten government was concentrated on the grasping financiers who collected the revenue.

Lavoisier entered this labyrinth of extortion and corruption with wide-open eyes and the best of intentions. The system was fundamentally bad, but it was all France had. It would have to be modified gradually if at all. Lavoisier resolved that his department at least would be run with honesty and efficiency. Of course he expected to make a good deal of money, but that was the psychology of the Old Régime, which regarded the government as a device for supporting the upper class. But what he did would be done in a nice way, with consideration for others as well as for himself.

The method of joining the Ferme was to buy a share of the total revenue from the *fermier général* for a lump sum which the *fermier général* paid over to the king. Lavoisier bought a third of a share for 520,000 livres and received the privilege of collecting the customs in certain western provinces. He plunged into the work with tremendous energy, not delegating the duties to subordinates as many of the other shareholders did. He even managed to combine the work with his constant scientific activities, filling the records of his business trips to the west of France with observations of mineral deposits and soil characteristics as well as with financial data. He reminded himself that all this money making was only a means to an end. He didn't forget that his real object was public service through scientific work unhampered by lack of financial means.

His colleagues in the Academy of Sciences looked on his new enterprise with a mixture of disapproval and envy. Some of them saw the catastrophe that was coming in 1789 and were sorry to see one of their number connected with the odious Ferme. Others resented the ease with which a man of capital and influence could increase his already ample fortune. They considered drawing up a formal protest, but thought better of it. The geometrician Fontaine remarked that Lavoisier's famous dinners would have more courses than ever, and the others consoled themselves with the thought that the more important

Lavoisier became in the government, the more favors
the Academy might expect to receive from it.

It soon became clear that Lavoisier's business interests were not going to interfere with his scientific work. Each minute of his day was accounted for,
and science got its full share. From six to nine in
the morning he worked in his laboratory, and from
seven to ten in the evening. He attended every meeting of the Academy and wrote numerous reports. A
tremendous correspondence kept a corps of secretaries busy. One whole day a week was devoted to
experiments which demanded continuous attention.
The Ferme Générale had to take the time left over.

The money which came to Lavoisier from the
Ferme and the energy with which he pushed his research made his laboratory a clearing house for
scientific information of every kind. All problems
which could not be solved because of the expense
were referred to him, and his apparatus and equipment were at the disposal of every scientist who had
none of his own. Problems which had remained unsolved for years were polished off in no time. The
members of the Academy decided that while the
Ferme remained the ranking evil in the government
of France, it had for once done the public a service
by providing Lavoisier with funds.

An amazing amount of work was performed in

Lavoisier's laboratory. Perhaps the first important accomplishment was the proof that water is not transmuted into earth by boiling, but others followed thick and fast. An experiment which had remained long undone because of expense was the proof that diamonds are made of carbon like ordinary charcoal and will burn to carbon dioxide if heated long enough. The little matter of destroying a number of costly diamonds did not stop Lavoisier for a moment. He heated them in all sorts of containers and proved that they eventually disappeared, leaving behind a gas which clouded lime water just as the gas from charcoal did.

By this time it was 1770 and Lavoisier already had a general idea of what his life work would be. It was no less than the destruction of the then dominant phlogiston theory. The root of Lavoisier's doubts about this theory was the fact that metals and other substances on combustion in air gain weight rather than lose it. To the phlogistians this seemed perfectly all right. Phlogiston, they believed, weighed less than nothing, and so of course a substance which gave it off gained weight in the process. But to Lavoisier this seemed an unwarrantable assumption. Nowhere else in Nature had he been able to find an example of negative weight, and his carefully quantitative methods had proved to him that what-

ever transformations matter may pass through, it
always remains exactly as heavy as before.

Very completely Lavoisier realized that such a
fundamental problem could not be solved offhand.
He would have to build up his new theory from the
bottom, repeating all the famous experiments of the
past, and checking carefully all the observations on
which the prevailing theories rested. Thanks to the
Ferme Générale, he was equipped for this tremendous task. He set to work.

Unlike Priestley, who never planned his work
carefully, or Cavendish, who considered science
merely an absorbing sport, Lavoisier went about his
task with a definite campaign mapped out. Each new
discovery, as soon as made, would fit in somewhere.
With the single exception of Boyle, who worked
along somewhat the same lines, Lavoisier was the
first chemist to see the problem as a whole. He resolved that in his work there would be no inspired
theories based on shaky facts, and no blind groping
after some tremendous but vague discovery. Nature,
he knew, was a vast and intricate machine whose individual parts moved according to a set of definite
laws. If he started with something absolutely sure
and absolutely definite and formed no conclusion
which wasn't in accordance with all the facts, he
knew he couldn't go wrong.

The bedrock law on which Lavoisier proposed to build his theory was the doctrine of the conservation of matter. This law had long been used by the more clear-headed chemists, but never sufficiently fortified with proofs. The alchemical notion that the elements could be transmuted into one another still persisted, and transmutation offered a safe refuge for the supporters of phlogiston who found themselves too hard-pressed. When confronted with some observation which did not fit into their theory, they would smile pleasantly and announce that the discrepancy of weight or material was accounted for by transmutation of one element into another.

The law of the conservation of matter can't be proved by positive means. The best we can do is demonstrate that it applies to each specific case. This was what Lavoisier did in his famous experiment of boiling water for a hundred and one days and proving that the solid matter left behind on evaporation came from the glass vessel, not from transmutation of water into earth. In this way he destroyed the pet proof of the transmutation chemists. He announced that as soon as they produced another, he'd destroy that too. No other was brought forward.

The next step was to repeat the famous experiments on which the phlogistians had built their theory. Lavoisier heated metals in air and weighed carefully both the metal and the oxide produced by calcination. He even went further and burned sulphur and phosphorus, collecting their combustion products and finding that in each case they weighed more than the unburned substance. It would seem an obvious conclusion to draw from this that something had been absorbed from the air used in combustion, but it wasn't as simple as we think it is now. The Eighteenth Century chemists had no reason to think that air was not a simple substance. It had always been regarded as such. Certainly no one had found a kind of air which was different from any other kind. Its properties all over the world were remarkably uniform, the slight variations being attributed quite correctly to the presence of water vapor. "And," said the phlogistians, "no matter how much metal or phosphorus you use, you can't absorb more than a fifth of a given quantity of air. How much more rational it is to say that the air remains free and combines with something from the metal, losing part of its volume in the process?"

Lavoisier was fully aware of this objection. At every point he was met by the question, "If air is absorbed, why isn't it all absorbed?" For the present he had nothing to answer. He hadn't been able to separate the air into active and non-active parts. We'll leave him in this dilemma and consider his other activities, one of which brought him the information his laboratory refused to provide.

All this time his business affairs had been going very well. Such had been his success with the tobacco monopoly, the customs, and a subsequent venture in salt, that the government entrusted him with the extremely important problem of saltpeter supply and made him *régisseur des poudres*. In Eighteenth Century warfare explosives were not used to the extent they are now, but gunpowder was, if possible, even more essential to the prosecution of successful war. It was practically the only munition of which an army had to have an unfailing supply. Weapons were simple and almost indestructible. Food could be collected from the surrounding country. An army didn't need gasoline, poison gas, or the thousand necessary supplies of modern warfare. But gunpowder it had to have, and this depended on the supply of saltpeter.

In the Eighteenth Century saltpeter was still "digged out of the bowels of the harmless earth," as Hotspur says in *Henry the Fourth,* but the process was not very satisfactory. It took a lot of digging and produced very little saltpeter. Lavoisier for the first time attacked the problem as a chemist, deciding that saltpeter came from decayed animal matter and developing methods for making it out of sewage,

*Portraits de M<sup>r</sup> & M<sup>me</sup> Lavoisier*
*d'après le tableau de David*

## LAVOISIER AND HIS WIFE
One of the greatest of all French scientists, Lavoisier
was condemned to death during the French Revolution with the words, "The Republic has no need for
learned men."

manure, and slaughter-house refuse instead of extracting it from the earth of barnyards and long-cultivated fields. He also found that the manufacturers had been treating it with leached wood ashes which contained very little of the necessary potassium. These two reforms and the economics effected by a drastic reorganization of the financial side of the business secured for France a much better and cheaper supply of gunpowder and raised Lavoisier into high favor with the government.

One of the reasons why Lavoisier had decided to go into financial politics was that he intended to develop a meeting place for all the natural philosophers of France and Europe. This would require both money and position. After he became controller of munitions such a center was secured, for he was given a residence and headquarters in the Little Arsenal on the Rue de la Cerisaie. Here he set up a well-equipped laboratory and issued standing invitations to all scientists to come and visit him when they felt so inclined. "The secrets of Nature," said Lavoisier, "can't be dragged into the light by one man. We must have the coöperation and the united efforts of many minds." He intended to meet and talk with every important philosopher. His dinners at the Little Arsenal became famous all over Europe.

At one of these dinners in 1774 Lavoisier met Priestley and heard from him an account of his discovery of oxygen. Priestley, faithful as always to the phlogiston theory, insisted on considering his gas mere dephlogisticated air, but Lavoisier knew better. It was just the information he'd been waiting for. It justified the expense of many a dinner.

This was a great step toward the completion of Lavoisier's new theory of chemistry. Although Priestley never admitted it, he had solved the problem which had blocked Lavoisier's campaign against phlogiston. It only needed a little careful laboratory work for Lavoisier to prove that a metal, mercury in this case, could take from the air the active gas responsible for combustion, leaving behind the inactive part which we now call nitrogen.

Now that this difficulty had been cleared up, Lavoisier could demolish the next stronghold of phlogiston, the regeneration of a metal from its calx by heating with powdered charcoal. It had long been known that when charcoal was burned in air it gave off a gas which clouded lime water. Lavoisier reasoned that if he should find such a gas in the vessel after heating a calx with charcoal, he could prove that the charcoal merely took from the calx the "active air" which the metal of the calx had in its turn extracted from the atmosphere. This proved to be the case, and Lavoisier had won another victory.

In all his scientific work Lavoisier was careful not to draw rash conclusions. "Time will tell," he said in one of his papers on combustion. "It is the fate of those who engage in physical and chemical experiments to see a new step to take as soon as the first has been taken. The road which has been presented to them appears to extend in proportion as they travel it." But after a triumph such as the demonstration that charcoal does not restore phlogiston to a calx, it was only human to generalize a bit. Lavoisier observed that many products of combustion may be dissolved in water to form acids. For once in his life he ran a little ahead of his facts and concluded that Priestley's dephlogisticated air was the "acid principle." So sure was he of this that he named it oxygen, the acid former. And he paid for his rashness by missing one of the greatest discoveries of the age, the synthesis of water. He burned inflammable air in oxygen in hopes of finding an acid like those formed by burning sulphur or phosphorus. He found no acid, and the small amount of moisture which was formed was concealed by the water in which he tried to collect the presumably gaseous acid. It wasn't until several years later that he heard of the work of Cavendish and learned what really had happened in his combustion tube.

By the time Lavoisier was thirty-five he was already a great figure in the social and political as well as the scientific affairs of France. If he had been a scientific hermit like Cavendish hiding in a deep burrow of research, he might have missed the feeling of impending calamity that hung over his country. But he was intimately in touch with affairs. Through the Ferme Générale he was practically a member of the government. And he knew that if something drastic weren't done, and quickly, there'd be an explosion such as France hadn't seen since the days of the Jacquerie.

Of course the fundamental trouble was the antiquated government and the privileges which it had granted for centuries to the parasite nobles and the decadent Church. It was a long story beginning far back in the Middle Ages. A tax here, a monopoly there. A grant of land to a noble, a grant of a toll-bridge franchise to an abbot. Reforms were attempted but never carried out, for they had to begin by abolishing some of the privileges of the nobles and clergy, and these classes through their influence with the king were always strong enough to defeat the reform.

On top of these abuses came the crushing burden of an extravagant court and almost constant warfare. Each year the Ferme Générale was called on to provide more money, and each year it had to apply more pressure to the wretched peasants who were the ultimate source of supply. But the money was never enough. The king and his corrupt favorites always spent more than came in, and vast debts accumulated which were repudiated from time to time to the ruin of the bankers who'd made loans to the crown.

One of the most alarming signs of trouble ahead was the desperation and sullen defiance of the peasants. Most of them were tenants on land belonging to the nobles or the Church, and they'd learned by bitter experience that the more they raised the more would be taken away by the landlord and the tax collector. If they improved the fertility of their fields by sound farming methods, their rent would be increased in proportion. So they lived from hand to mouth in grumbling hopelessness. They knew they'd be allowed just enough to keep body and soul together. They didn't intend to produce a surplus for the use of their oppressors.

Ever since Lavoisier had first joined the Ferme Générale, he had realized that things were in a pretty bad way. He had been able to put through a few minor reforms, notably in the administration of the tobacco and saltpeter monopolies, but in 1778 he embarked on a more far-reaching experiment. He bought the large estate of Frechines near Blois to develop into a model farm. He had observed during his travels about the country that almost nowhere was the land producing as much as it would if properly cultivated, and he thought he knew the reason why. The peasants were not stupid—decidedly not. They were merely ignorant and discouraged—ignorant because no one had taken the trouble to give them up-to-date information, and discouraged because they knew that if they improved their methods, the resulting benefits would immediately be taken from them.

Lavoisier was confident that if the peasants knew that their efforts would be well rewarded they would improve their own condition and that of the government which depended on them. But to argue along this line in government circles was useless. To the nobles and clergy who owned the land and the tax farmers who collected the revenue, the peasant was a beast of burden to be loaded with as much as he could carry. That he would carry more if less heavily loaded was a parodox beyond the comprehension of the blind or cynical high officials. So Lavoisier decided to start at the other end.

The first thing he did at Frechines was to assure his peasants that they would receive the benefits of any improvements they made. He promised not to raise their rents as the productivity of their fields increased, and he promised to use his position in the Ferme Générale to protect them from the rapacity of the tax collectors. If they followed his advice about crop rotation and fertilization, they'd raise better crops and these would belong to them alone. For his pay he'd take the chance to show the government what could be accomplished by treating them justly.

It was slow work. Although Lavoisier held only a very minor title, he was an aristocrat in the eyes of his peasants, and they looked on all aristocrats with well-founded suspicion. It was hard to convince them that he was sincere and doubly hard to make them adopt farming methods which were strange and new. To gain their confidence, Lavoisier started with small plots of the poorest land, improving them gradually until they equaled and surpassed the best land on the estate. As he grew in favor with his people, he introduced more reforms. He was among the first to plant potatoes. He imported cattle from Spain and milch cows from Chanteloup. He distributed improved seed. Soon the wheat yield had doubled and the number of cattle had multiplied by five. The peasants were prosperous and grateful. To them Lavoisier was a father, almost a saint. Each trip he made to his

estate sent him back to Paris more thankful that
he'd tried the experiment.

> "Such an investment [said he] does not
> present the brilliant speculation of stock buying
> or gambling in the public bills-of-exchange, but
> neither is it accompanied with the same risks and
> the same reverses. One's success causes no tears;
> on the contrary it is attended by the blessings of
> the poor. An intelligent proprietor cannot make
> his farm increase in value with improvements
> without spreading about him comfort and hap-
> piness. A rich and abundant vegetation, a numer-
> ous population, a picture of prosperity, these
> are the rewards for his pains."

Sometimes Madame Lavoisier went to the farm
to watch her husband among his people.

> "One should see him [she wrote] in that
> house in the society of the villagers, as mag-
> istrate of the peace reëstablishing harmony be-
> tween neighbors, giving an example of all the
> patriarchal virtues, caring for the sick not
> merely with money but by his visits and per-
> sonal attentions, encouraging them to be patient
> and hopeful, and founding a school for the gen-
> eration which, before his coming, was growing
> up without culture."

Lavoisier's success in scientific farming was so
great that in 1785 he was put in charge of a com-

mittee on agriculture which the government was
finally persuaded to appoint. It had large powers
over such things as flax manufacture and the duties on
agricultural products, but as soon as it made any
real attempts to improve the condition of the peasants it ran head on against the ancient privileges of
nobility and clergy, against the forced labor laws
inherited from feudal times, against the *gabelle,* the
*taille,* and the *corvée.* Lavoisier saw that the trouble
lay deeper than any committee on agriculture could
hope to reach.

In a report to the Controller General he put his
finger on the basic fault of his country's despotic and
antiquated government:

"We know [he wrote] that the true end of a
government should be to increase the sum of en-
joyment, happiness, and well-being of all individuals. If commerce has been more encouraged, more protected than agriculture, it is
because the profession of the merchant is practised by a class of citizens of a more awakened
order who know how to talk and write, who
live in cities, who make there a group whose
voice is easily heard. The unfortunate farmer
groans in his thatched cottage; he has neither
representative nor defender, and his interests
have not been counted of any value in the dis-

tribution which has been made of departments
for the administration of the kingdom."

Lavoisier was not the man to stop with generaliza-
tions. He went on to list in detail all the burdens
which the peasant had to bear—the ecclesiastical
levies, the unjust and often capricious taxation, the
forced labor, and the compulsory use of the landlord's grist mill. All these would have to be abolished if the peasants were to be restored to prosperity.

By this time it was 1788, and conditions had
become so bad that the government was thoroughly
alarmed. Almost daily reports of disorder reached
the capital. Sometimes it was the peasants rebelling
against their feudal lords, sometimes a manor house
set on fire in the night, sometimes a tax collector
found murdered by the roadside, sometimes a mob
of a provincial city welling out of the slums to burn
and destroy. At Paris, Necker, the able Minister of
Finance, was almost at his wit's end. Every reform
move he made was blocked by the parasites cluster-
ing around the throne. Every change in administration was opposed by some selfish interest which bene-
fited by things as they were. And the king played with
his locksmith tools, hardly conscious that there was
trouble.

Finally Necker decided that if he couldn't accomplish anything with the court, he could at least set up governing bodies which would function with more justice and energy. As an experiment in decentralization, he called the Provincial Assembly of Orléans, which consisted of twenty-five members appointed by the king and an equal number elected by the appointees. Lavoisier was technically a nobleman, but he was appointed as a commoner to represent the district of Romorantin near his model farm. At the first session in September he acted as secretary; at the second in November his influence was dominant.

In spite of the fact that it was not elective but appointed by the king, the assembly of Orléans was remarkably liberal. The reforms it proposed sound advanced even in the Twentieth Century. Besides an efficient system for peasant education, Lavoisier worked out a scheme of insurance designed to assure the peasant of a secure old age if he cultivated his land effectively during his active period. He made strenuous efforts to abolish forced labor and feudal dues and submitted a report on general taxation. If the Provincial Assembly of Orléans inspired by Lavoisier had been allowed to finish its work, the Revolution might never have occurred. But it was too late. The central government was bankrupt. Revolts were threatening in every corner of the kingdom. Necker in desperation decided to call the Estates General for the first time in 175 years. The Fall of the Bastille was only a few weeks ahead. Few realized it, but the Revolution had already begun.

The calling of the Estates General was the cause of universal rejoicing. Full of confidence and hope, Lavoisier went to Paris to join it as the delegate from Blois. "Now," he thought, "we have all the nation gathered together in assembly. We can wipe the slate clean and start afresh with justice for everyone from peasant to king." But it wasn't so simple. Not everyone was as public-spirited as Lavoisier. The great nobles, the higher clergy, and the bankers who had fattened on the graft of the Ferme Générale were not going to give up their privileges without a struggle. The Estates might meet, but they would find plenty of obstacles in their way.

The French Revolution was far from a simple thing. It didn't take place in a day or a year. As the power slipped from the paralytic hand of the king and favorites, each other class made an attempt to secure it. First the Liberal nobles; then the bourgeoisie, then the representatives of the people, and finally the mob from the gutters of Paris who ruled with ignorant ferocity, striking down friend and foe alike. For a while it looked as if the Estates General led by enlightened noblemen like Lafayette and

Mirabeau would keep the upper hand, but trouble soon developed. In the first place there was a good deal of confusion about the relative powers of the three estates, the nobles, clergy, and commons. If they sat as separate houses each with a veto power over the acts of the other, the privileged classes which controlled two of them would be able to block all reforms. Whereas, if they sat together, the commons who equaled in numbers both the other houses together would be able with the aid of Liberal nobles and clergymen to out-vote their reactionary opponents. There was a long wrangle with charges and counter-charges until finally the king ordered the reluctant upper houses to sit with the commons.

This was a victory for the Liberals, but their rejoicing was short, for the Parisian mob grew more and more threatening. The government began to concentrate what troops it could gather near the capital. Paris defied them. Barricades were built over night. And finally on July 14, 1789, the first of many storms broke. On a rumor that the king was going to dissolve the Assembly, the mob of Paris broke loose and destroyed the ancient prison of the Bastille, the symbol of royal tyranny. Wholly unchecked by royal or other authority, it tore through the city seizing all aristocrats it could lay its hands on and hanging them to the lamp posts—Lavoisier's lamp posts, for it was only some twenty years since he had designed them and they were made to last. July 14th was a taste of what was coming. The great aristocrats began to emigrate to foreign countries. Only the foolish and the idealistic stayed. Among the latter was Lavoisier, whose conscience wouldn't let him desert the country he'd served so long and so well.

One of the surprising things about the French Revolution was the slowness with which it took place. After July 14th the Parisian mob was always a threatening cloud in the background and in many of the provinces the peasants had risen and reduced their neighborhood to anarchy. But still the saner leaders of the Estates clung to hope and continued their deliberations at Versailles. Concession after concession they wrung from the terrified upper classes. Feudal dues were renounced. Titles were abolished. The rights of man were proclaimed, and an admirable constitution was adopted. But things had gone too far. The Parisian mob broke loose again and again, each time with greater ferocity. Demagogues rose from the gutters to take the places of the moderate leaders who were trying to avert the catastrophe. Most of the aristocrats had fled, but Lavoisier stayed on, making no attempt to save his skin. He had done much to start the Revolution. He felt that he should see it through.

In all the confusion and bloodshed of those terrible years there was one place where he could find peace and contentment. This was the Little Arsenal where his laboratory was still in working order. The Birmingham mob had destroyed Priestley's apparatus, but the mob of Paris was of a different temper. It did not attack science as such. So in his laboratory Lavoisier could find the peace he needed to finish his work.

Perhaps he saw the end drawing near. At any rate he now began to gather his discoveries into a complete theory of chemistry. He resolved that if the world he knew should fall into chaos, he would leave at least one useful legacy for future generations. This was his famous *Traité élémentaire de chimie,* the first modern textbook of chemistry. It was the death warrant of the phlogiston theory. Carefully, completely, and accurately, Lavoisier explained his new conceptions of chemical law. He demonstrated with convincing experiments how the air contains oxygen and nitrogen, how the oxygen combines with a burning substance, how water contains oxygen and hydrogen, how all these reactions take place according to exact laws, and how this proves that chemistry is a definite mathematical science.

The *Traité* was at once hailed as the most important contribution to chemistry ever published.

The members of the Academy, those who hadn't fled the country, were unanimous. Foreign scientists wrote to Lavoisier in glowing terms. It was translated at once into English, German, Spanish, and Italian. All Europe proclaimed it a masterpiece, a great forward step toward the scientific age to come.

It was a strange time to publish a masterpiece, and to Lavoisier sitting in his laboratory at the Little Arsenal the chorus of praise must have sounded rather like a last tribute. It didn't take much penetration to tell him that his career was nearly over. Already he knew what it was like to have a mob howling for his blood. To the scientists of Europe he was Lavoisier the chemist, but to the mob he was Lavoisier of the Ferme Générale, and anyone who had ever been connected with the hated Ferme was marked down on the mob's darkest blacklist. As early as 1789 he got a hint of how the populace felt toward him. There'd been an explosion in one of his powder mills in which two men were killed. This had been given great publicity by the enemies of the munitions administrators, but the feeling against them did not come to a head until August 6th when Lavoisier decided to move some of the powder from the crowded arsenal to storehouses at Rouen and Metz. The rumor spread that he was selling it to the enemy, and at once the mob was at his door. The riot was suppressed, but Lavoisier knew that his troubles were not over.

One of the alarming signs of the course events were taking was the rise to influence of Marat, the bloodthirsty leader of the Paris slums. To Marat revolution did not mean an orderly reform of government or an abolition of abuses. He rejected such mild measures as inconclusive. All aristocrats, all rich men, all army officers, all officials of the Old Régime must be destroyed, he thought, before liberty could be attained. In his paper, "The Friend of the People," he preached murder, mutiny, and rape. He attended the National Assembly with a brace of pistols at his belt and marched at the head of every mob which promised blood for him to enjoy.

Now Marat had started his career as a medical student, but finding the course of study too strenuous, had taken a short cut and begun selling quack medicines on the streets of Paris. He also rather fancied himself as a chemist and published a treatise on fire which he announced had been praised by the Academy of Science. The Academy hadn't even seen it, far less praised it. Lavoisier pointed out several glaring errors in the book and dismissed it in a few disdainful words, thus gaining the eternal hatred of Marat, who never forgot an enemy.

More than twenty years had passed since this oc-

curred. Lavoisier had forgotten all about it, but
Marat had not. He bided his time until 1791. Then
he launched a savage attack on Lavoisier:

> "I denounce this Corypheus of the Charla-
> tans, Sieur Lavoisier, son of a land-grabber,
> chemical apprentice, pupil of the Genevese
> stock-jobber, *fermier-général, regisseur* of pow-
> der and saltpeter, administrator of the Discount
> Bank, secretary of the king, member of the
> Academy of Science. Would to heaven that he
> had been strung to the lamp post on August 6th.
> The electors of La Culture would then not have
> to blush for having nominated him."

Nor was this all. As a member of the Ferme
Générale he was attacked in the journals, in the all-
powerful political clubs, and in the National Assembly itself. One by one his offices were taken from
him. He was expelled from the Assembly. He was
deposed as *regisseur des poudres*. He was even driven
from the Commission of Weights and Measures. By
this time Lavoisier had long since given up any hope
of saving his own skin. He knew he was already as
good as dead. But when the revolutionary extremists
attacked his beloved Academy of Science he made
one last fight with his old vigor.

The revolutionists hated the Academy for the
simple reason that it was a royal institution, but it

was hard to find any specific charges against its members. They were mostly elderly scientists who had taken no part in public affairs. But when the Revolution had reached this stage, specific charges were not needed. The "friends of the people" had invented the blanket crime of "incivism" to cover all such cases. It meant anything and everything from mere possession of property to high treason. In 1792 the Assembly demanded that the Academy expel all "incivic" members. Lavoisier and the other officers of the society of course protested that the Academy had nothing to do with the political opinions of its members. They managed to stave off the blow until the summer of 1793, when the Academy was finally abolished.

Lavoisier had hoped to save the Academy. He had no hope of saving himself. The Year of the Terror was approaching, and he was forever stigmatized as having belonged to the hated Ferme Générale. But to the end he worked as he had worked all his life, improving the manufacture of saltpeter, even assisting the Commission of Weights and Measures from which he had been expelled. Each day he expected to be his last. His enemy Marat had died under the dagger of Charlotte Corday, but he had plenty of others. He had been a *fermier*. That was enough.

On November 24, 1793, the Assembly decreed the

arrest of all members of the Ferme Générale who had signed the leases of David, De Salzard, and De Meyer. Among these was Lavoisier. He was taken to the convent of Port-Royal, then used as a prison, to await the forgone conclusion of his trial.

On the fifth of May the prisoners were ordered to appear before the revolutionary tribunal. This was the end. The tribunal had an almost hundred per cent perfect record. To be sent before it was the equivalent of a death sentence. The specific charge against Lavoisier was adulteration of the people's tobacco, but this was only mentioned in passing. A final appeal was made for him. His patriotic services were recounted—the improved manufacture of gunpowder which had made possible the Republic's victories, the work for the peasants of Frechines, the numberless reports on reordering the kingdom's economic system, his great accomplishments in science. But it was useless. The terrible Coffinhals presided and he had no interest in such matters. He hardly listened to the appeal. When the speech came to an end, he looked up and pronounced the final words:

"The Republic has no need for learned men. Let Justice take its course."

The next day the prisoners were led to the Place de la Revolution, guillotined without ceremony, and their bodies thrown into an unmarked grave in the cemetery of Madeleine. France had executed its most distinguished scientist. As one of the witnesses said the next day, "It took only a moment to cut off that head. It will take a hundred years to produce another like it."

Lavoisier's execution was the work of the Reign of Terror at its height. After a few more bloody weeks France awoke as if from a nightmare. Robespierre followed his victims to the scaffold and with him went Coffinhals and a hundred others. The Terror was over. Even before the reaction was fully under way, the leaders of popular opinion realized the mistake they had made. Protests poured in from all over France, all over Europe, for Lavoisier was revered by every lover of knowledge. Even the people, now that sanity had returned, remembered his unselfish work at Frechines, his devotion to the public good. Only a few months after his death, the National Assembly voted to restore his confiscated property to his widow, and one of his bitterest persecutors, Fourcroy, thought it best to compose an elaborate *éloge* to his memory.

Lavoisier died at the height of his productive period, but if on the way to the scaffold he looked back over his career, he could congratulate himself on having finished at least one job. He'd found the science of chemistry in utter confusion, wandering in a labyrinth of blind alleys, following the false god phlogiston. He left it with a clear road to follow. Before Lavoisier, the chemist was a theorist arguing vaguely about reactions he did not understand. After Lavoisier he was a searcher for the definite laws of Nature. Lavoisier had shown the way.

The Harvest of Peaceful Middle Age

# CHAPTER X

## The Harvest of Peaceful Middle Age

SCIENCES have a life cycle. They grow up, come of age, and some of them die. Chemistry is a science which has passed through all its growing periods, attained maturity, and is now ready, like an old pear tree, to produce a generous crop of fruit each year without getting any larger or attempting new things.

It isn't the fault of the chemists that this change of pace has occurred. It is because chemistry, more than other sciences, is limited by definition. Its province is the molecule and its transformations. It does not penetrate into the atom. That is physics. It does not pry into the living cell. That is biology. There's an invisible line beyond which a modern chemist cannot go without becoming a physicist or a biologist. The chemists are like a group of settlers on a large and fertile island who have pushed their frontier to the sea on all sides and are left with the choice of migrating to other lands or staying at home to develop the ground they have won.

Lavoisier died in 1794, Priestley in 1804, and Cavendish in 1810. With their passing chemistry may be said to have come of age. Its youthful battles were over. The chemists of the Nineteenth Century did not have to contend with theology or the other intellectual legacies of the Middle Ages. The attitude of the Nineteenth Century toward science was substantially the attitude of the present. The scientists were honored by the public, subsidized by the state, and let alone by the Church. They could work in peace to build up the wonderful if somewhat terrifying thing we call modern civilization.

But with the struggle went some of the glory also. The later chemists were not enemies of society like Roger Bacon, outcasts like Paracelsus, or reluctant inconoclasts like Boyle and Priestley, half afraid of what they were starting. They did not have to push boldly out into the unknown. Their road had plenty of sign posts. Lavoisier had put them there. All that the Nineteenth Century chemists had to do was to follow his directions and they couldn't go wrong.

After Lavoisier the course of true chemistry ran remarkably smoothly. Discoveries came thick and fast—like the olives from a bottle when the first reluctant one has been dislodged. Men knew how the trick was done, and they proceeded to do it. There

were many workers now instead of only a handful—
so many that the credit for each discovery has to be
divided between a dozen claimants. And as the
science grew, it divided into special branches. There
were no more old-fashioned natural philosophers
who covered the whole field of knowledge. The
breadth of interest which made the early chemists so
fascinating was no longer possible in a world grown
many-sided and complicated. Lavoisier touched
every human concern from politics to theoretical
chemistry, but after his death this could no longer
be done. The field was too large. Each specialty had
its corps of devoted workers. Dalton, Gay-Lussac,
Avogadro, and Kekulé discovered the molecule and
the laws which govern its formation. Davy, Fara-
day, and Arrhenius determined the part which electricity plays in chemical reaction. Liebig, Wöhler,
Dumas, and Kopp explored the labyrinth of organic
chemistry. The scientific age had begun in earnest.
Instead of a few courageous pioneers working against
violent opposition to establish the fundamental laws,
a host of specialists, secure in the universities and
state-subsidized laboratories, gathered the harvest
the pioneers had sown.

At the present time chemistry can hardly be called
"pure" science. It is largely technology—the investi-

gation of specific problems with a practical object in view. New dyes, new plastics, new finishes—these are the concerns of the modern laboratory. The chemical engineer touches every human activity from farming to warfare, and civilization could not continue without him. But it's a long way from the pioneers with their wide-ranging hopes and plans to the modern chemist trying to make an acceptable soap out of some evil-smelling oil from the tropics.

If we want to find men like Boyle and Lavoisier whose thoughts reach out into the unknown regions beyond present knowledge, we have to leave chemistry and follow the frontier of science into physics, mathematics, biology, and psychology. Here we'll find unsolved problems. Hints of intellectual revolutions to come which terrify the modern mind as the first glimpses of science terrified the bishops of Charles the Second. It is still too early to judge, but Einstein may overthrow the conventional physics as completely as Lavoisier overthrew phlogistonism. Relativity, the quantum theory, and the startling developments of biology and psychology may cause a modification of human life as far-reaching as the industrial revolution which followed the development of Eighteenth Century science. We are still living on the fruits of Eighteenth Century thought.

There has been no radical change in intellectual viewpoint since the firm establishment of the scientific method. But there are signs that another revolution is due.

THE END

# Favorite Authors and Their Books

THERE are thousands of Americans of all ages who place that prince of adventurers, Richard Halliburton, at the top of their list of favorite authors. Mr. Halliburton has traveled in remote places far from tourist routes and after each of his trips he has returned to the United States and written a book of gay, romantic adventure. These books—*The Royal Road to Romance; The Glorious Adventure; New Worlds to Conquer; The Flying Carpet; Seven League Boots;*—are all in the Star Library. Such is their popularity that it has been the experience of the publishers that anyone who reads one of them is not satisfied until he has read them all.

In another field of literature Stefan Zweig is internationally famous, each of his brilliant biographies having earned the highest praise. We hesitate to state our own preference but are proud to publish both Mr. Zweig's MARIE ANTOINETTE and his MARY, QUEEN OF SCOTLAND AND THE ISLES in the Star Library. These are books which lend distinction and prestige to a publisher's list as well as to a personal library. Many people prefer biographies to all other reading. In addition to Stefan Zweig's books, the Star Library contains such titles as *Napoleon* by Emil Ludwig, *Henry VIII* by Francis Hackett, Vincent Sheean's absorbing autobiography, *Personal History,* and *The Life and Times of Rembrandt,* one of several immensely popular books by Hendrik van Loon included in this series.

In view of the fact that the Star Library contains books by so many "favorite authors," many discerning readers have instructed their bookseller to inform them when a fresh title joins this list. By doing so they save themselves time and trouble and if they decide to make a purchase they add a volume to their own library which will be read by the whole family.